Memoirs of the American Mathematical Society

Number 240

Tai-Ping Liu

Admissible solutions of hyperbolic conservation laws

Published by the
AMERICAN MATHEMATICAL SOCIETY

Providence, Rhode Island, USA

March 1981 · Volume 30 · Number 240 (first of 3 numbers)

<u>AMS(MOS) subject classification (1980)</u>. Primary 76L05; 35L65
                                               Secondary 35B40; 35A40

<u>Key words and phrases</u>. Hyperbolic conservation laws, admissibility crite-
                          rion, Riemann problem, random choice method, wave
                          partition technique.

Library of Congress Cataloging in Publication Data

Liu, Tai-Ping, 1945-
    Admissible solutions of hyperbolic conservation laws.

    (Memoirs of the American Mathematical Society ;
no. 240 ISSN 0065-9266)
    "Volume 30 ... (first Of 3 numbers)."
    Bibliography: p.
    1. Shock waves. 2. Conservation laws (Physics)
3. Differential equations, Hyperbolic--Numerical
solutions. I. Title. II. Series: American Mathematical
Society. Memoirs ; no. 240.
QA3.A57 no. 240   [QA921]   510s   [531'.1133]
ISBN 0-8218-2240-3   80-28506

(2.1) $\sigma'[u] = (\frac{\partial f}{\partial u} - \sigma)u'$, $u' \equiv \frac{du}{d\mu_i}$, $[u] \equiv u - u_0$,   etc.

which yields (i). Since r_i is the i-th right eigenvector of $\partial f/\partial u$, we have from (2.1) that

$$(\frac{\partial f}{\partial u} - \sigma)(u' - r_i) = \sigma'[u] - (\lambda_i - \sigma)r_i .$$

We differentiate the above to obtain

(2.2) $(\frac{\partial f}{\partial u} - \sigma)'(u' - r_i) + (\frac{\partial f}{\partial u} - \sigma)(u'' - r_i')$

$$= \sigma''[u] + \sigma'u' - (\lambda_i' - \sigma')r_i - (\lambda_i - \sigma)r_i' .$$

From (i), the above equality yields

$$(\frac{\partial f}{\partial u} - \sigma)(u'' - r_i') = (2\sigma' - \lambda_i')r_i \quad \text{at} \quad u = u_0 .$$

Since u' and r_i are unit vectors, $u'' - r_i'$ is orthogonal to $u' = r_i$ at $u = u_0$. Moreover $\sigma = \lambda_i$ at $u = u_0$ and so $(A - \sigma)(u'' - r_i')$ is a linear combination of $\{r_j = j \neq 2\}$ at $u = u_0$. Thus (ii), (iii) follow from the last identity.   Q.E.D.

LEMMA 2.3. Suppose that μ_i is any nonsingular parameter along $S_i(u_0)$ and increases in the direction of $r_i(u)$. Then for any u on $S_i(u_0)$ and in a small neighborhood of u_0.

   (i) $\lambda_i(u) > \sigma(u_0, u)$ (or $\lambda_i(u) < \sigma(u_0, u)$) if and only if

$$\frac{d\sigma(u_0, u)}{d\mu_i} > 0 \quad (\text{or} \quad \frac{d\sigma(u_0, u)}{d\mu_i} < 0),$$

   (ii) $S_i(u_0)$ is tangent to $R_i(u)$ at u if $\sigma(u_0, u) = \lambda_i(u)$.

   PROOF: For $u \in S_i(u_0)$, we write

(2.3) $u - u_0 = \sum_{j=1}^{n} \alpha_j r_j(u)$, $\alpha_j \equiv \alpha_j(u_0, u)$,

(2.4) $\frac{du}{d\mu_i} = \sum_{j=1}^{n} \beta_j r_j(u)$, $\beta_j \equiv \beta_j(u_0, u)$.

From (i) of Lemma 2.2, we see that

It is clear from (5.2) and (5.5) that

$$0 \leq \tilde{\sigma} - \sigma_2 \leq \sigma_1 - \sigma_2 \equiv \theta.$$

We have from (5.8) and the above two estimates that

$$\tilde{\sigma}(u_2 - u_0) - [f(u_2) - f(u_0)] = O(1)\beta\theta \min(\alpha_1, \alpha_2).$$

It follows from Lemma 2.5 that (ii) holds and

(5.9) $$|\tilde{\sigma} - \sigma| = O(1)\theta \min(\alpha_1, \alpha_2).$$

The above estimate and (5.5) imply (iii).

It remains to verify that (u_0, u_*) is admissible. It is clear from (5.5) and (5.9) that for weak waves

(5.10) $$\sigma_2 \leq \sigma \leq \sigma_1.$$

Since (u_0, u_1) is admissible, it follows from (5.10) that if (u_0, u_*) is not admissible then there exists $\tilde{u} \in S_i(u_0)$, $u_1^i \leq \tilde{u}^i < u_*^i \equiv u_2^i$, such that $\sigma(u_0, \tilde{u}) = \sigma(u_0, u_*) \equiv \sigma$. It is an easy consequence of the jump condition (R-H) that this implies $\tilde{u} \in S_i(u_*)$ and $\sigma(u_*, \tilde{u}) = \sigma(u_0, u_*) \equiv \sigma$. Hence we have from (5.9) that

(5.11) $$\sigma(u_*, \tilde{u}) - \tilde{\sigma} = O(1)\theta \min(\alpha_1, \alpha_2).$$

On the other hand, we know that (u_1, u_2) is admissible and thus

(5.12) $$\sigma_2 \geq \sigma(u_2, u)$$

for any u on $S_i(u_2)$ between u_1 and u_2. It follows from (ii) of the lemma we just proved that we may choose u in (5.12) so that $\|u - \tilde{u}\| = O(1)\,\theta \min(\alpha_1, \alpha_2)$ so that (5.11) and (5.12) yield

(5.13) $$\sigma_2 - \tilde{\sigma} \geq O(1)\theta \min(\alpha_1, \alpha_2).$$

Since $\theta = \sigma_1 - \sigma_2 \geq 0$ and $\sigma_1 \geq \tilde{\sigma} \geq \sigma_2$ (c.f. (5.5)), we may assume, without loss of generality, that $\tilde{\sigma} - \sigma_2 \geq \frac{\theta}{2}$. This would contradict (5.13) when $\theta > 0$ because we assume that either $O(1)$ in (5.13) or the strengths α_1 and α_2 of waves are small. When $\theta = 0$, it is clear from (R-H) that $u_* = u_2$ and (u_1, u_2) is admissible. We have thus shown that (u_0, u_*) is admissible. This completes the proof of the lemma.   Q.E.D.

(i) $\qquad \sum\limits_{k=1}^{\ell} \| w_k - u_{j_k} \| = O(\varepsilon),$

and, for any k even and $1 \le k \le \ell$, there exists a monotone sequence $\{\bar{u}_j : j_{k-1} \le j \le j_k\}$ on $R_i(w_{k-1}) = R_i(w_k)$ with $\bar{u}_{j_{n-1}} = w_{k-1}$, $\bar{u}_{j_k} = w_k$ and

(ii) $\qquad \| \bar{u}_j - u_j \| = O(\varepsilon), \qquad j_{k-1} \le j \le j_k,$

and, for any simple discontinuity (u_{k-1}, u_k), k odd and $1 \le k \le \ell$,

(iii) $\qquad j_{k-1} = j_k - 1$

and $(u_{j_{k-1}}, u_{j_k})$ is an i-discontinuity; or there exists a monotone sequence $j^1 = j_{k-1}, j^2, \ldots, j^s = j_k$ such that $\{(u_{j_\tau-1}, u_{j_\tau}),$ τ even, $1 \le \tau \le s\}$, are i-discontinuities and $u_{j_\tau-1}$ and u_{j_τ}, τ odd, are related by i-rarefaction waves (mod ε) in the sense of (i) above. The last case above is possible only if (w_{k-1}, w_k) is a composite of $(\bar{w}_1, \bar{w}_2)$, $(\bar{w}_2, \bar{w}_3), \ldots, (\bar{w}_{s-1}, \bar{w}_s)$. In this case, we have

(iv) $\qquad \sum\limits_{\tau=1}^{s} \| \bar{w}_\tau - u_{j\tau} \| = O(\varepsilon).$

Here, as usual, $O(\varepsilon) \to 0$ as $\varepsilon \to 0$.

PROOF: We consider the following three cases:

Case 1: Every state u between (u_{j-1}, u_j), $j = 1, 2, \ldots, m$, lies in $\Omega^i_{2\varepsilon^{1/3}}$.

Case 2: There exists an i-rarefaction wave (u_{j-1}, u_j) with the property that $u \notin \Omega^i_{2\varepsilon^{1/3}}$ for some u between u_{j-1} and u_j; and there does not exist $\bar{u}$ and $\bar{\bar{u}}$ between some i-discontinuity (u_{j-1}, u_j) such that $\bar{u} \in \Omega^i_{\varepsilon^{1/3}}$ and $\bar{\bar{u}} \notin \Omega^i_{2\varepsilon^{1/3}}$.

Case 3: There exist $\bar{u}$ and $\bar{\bar{u}}$ between some i-discontinuity (u_{j-1}, u_j) such that $\bar{u} \in \Omega^i_{\varepsilon^{1/3}}$ and $\bar{\bar{u}} \notin \Omega^i_{2\varepsilon^{1/3}}$.

In case 1, it follows from (ii) of Lemma 2.4 that all states u between (u_{j-1}, u_j), $j = 1, 2, \ldots, m$, take values along $R_i(u_0) + E$ where the error term E satisfies (c.f. (3.2))

$$E = O(1) \varepsilon \sum\limits_{j=1}^{m} |u_j - u_{j-1}|^2 = O(1) \varepsilon M\phi_i(2\varepsilon^{1/3}).$$

$\bar{\theta} \equiv \sum \{\theta_\ell(u,u') + \theta_r(u,u') : (u,u')$ any discontinuity crossing $J_1 \cap J_2$
between (u_0,u_2) and $(u_-,u_+)\}$.

It is clear that $\theta_r(u_0,u_2) = \theta_r(u_1,u_2) + \sigma(u_0,u_2) - \sigma(u_1,u_2)$. Direct calculations using these identities yield

$$\gamma(\alpha_1+\alpha_2)\theta(u_0,u_2;u_-,u_+) = \gamma\alpha_1\theta(u_0,u_1;u_-,u_+) + \gamma\alpha_2\theta(u_1,u_2;u_-,u_+).$$

In other words, the potential amount of wave interaction between (u_0,u_1) and (u_-,u_+) and between (u_1,u_2) and (u_-,u_+) equals that between (u_1,u_2) and (u_-,u_+). On the other hand, (u_0,u_1) and (u_1,u_2) interact on J_1 but they are not interacting on J_2. Consequently, $Q(J_2) - Q(J_1)$ $=-Q(\Delta)$ and thus $F(J_2) - F(J_1) = -KQ(\Delta)$. In the above analysis we have suppressed the nonlinear effect; in general we have $F(J_2) - F(J_1) \le$ $-KQ(\Delta) + O(1)Q(\Delta)$.

THEOREM 8.1. Suppose that the initial data $u(x,0)$ has bounded total variation T.V. and either T.V. or $\max\|f''\|$ is sufficiently small. Then there exists a suitable positive constant K such that the nonlinear functional $F(J)$ is uniformly bounded for all I-wave J, and

$$F(J) = O(1)T.V.$$

PROOF: We will prove by induction that

$$F(J_2) \le F(J_1)$$

for any I-curves J_1 and J_2, J_2 an immediate successor of J_1. For the I-curve O in $0 \le t \le s$, it is clear that

$$F(O) = T.V. + O(1)(T.V.)^2$$

which is less than 2 T.V. if $O(1)$T.V. is sufficiently small. The induction hypothesis is that

$$F(J_1) \le 2 \text{ T.V.}$$

Suppose that Δ is the diamond between J_1 and J_2 and the waves entering Δ is the solutions of Riemann problems (u_ℓ,u_m) and (u_m,u_r). For brevity, we will denote by $Q_s(\Delta) \equiv Q_s(u_\ell,u_m,u_r)$, etc. From Theorem 7.3 we have

$$L(J_2) - L(J_1) \le O(1)Q(\Delta).$$

$$\sum_{\ell=1}^{N} w_k^+(T_1)\Big|_{I_\ell} \geq w_k^+(T_1) - \varepsilon.$$

$$\left|\sum_{\ell=1}^{N} w_k^-(T_1)\Big|_{I_\ell}\right| \leq \varepsilon.$$

The above partition is possible because $dw_k^+(T_1)$ is a Borel measure. Since $\{dw_k^+(u_{r_{ij}}, (x,T_1) + dw_k^-(u_{r_{ij}}(x,T_1))\}$ tends to $d\overline{w}_k^+(T_1) + d\overline{w}_k^-(T_1) = dw_k^+(T_1) + dw_k^-(T_1)$, the above estimate implies that there exists $J_1 > 0$ and intervals $I_\ell(u_{r_{ij}})$, $\ell = 1,2,\ldots,N = N(\varepsilon)$, such that for $ij > J_1$ the total amount of k-expansion waves in $u_{r_{ij}}$ over $I_\ell(u_{r_{ij}})$, $\ell = 1,2,\ldots,N$, is larger than $w_k^+(T_1) - 2\varepsilon$:

$$(11.1) \qquad \sum_{\ell=1}^{N} w_k^+(T_1;u_{r_{ij}})\Big|_{I_\ell(u_{r_{ij}})} + \sum_{\ell=1}^{N} w_k^-(T_1;u_{r_{ij}})\Big|_{I_\ell(u_{r_{ij}})}$$

$$\geq w_k^+(T_1) - 3\varepsilon, \qquad\qquad\qquad\qquad ij > J_1.$$

We now estimate $w_k^+(T_1;u_{r_{ij}})\Big|_{I_\ell(u_{r_{ij}})}$, $\ell = 1,2,\ldots,N$. Let $\chi_{k,\ell}^1(u_{r_{ij}})$ $(\chi_{k,\ell}^2(u_{r_{ij}}))$ be the first (last) k-characteristic of type I (c.f. Section 9) from the left which passes through $I_\ell(u_{r_{ij}})$ at time T_1. The region between time $= T_0$, time $= T_1$, $\chi_{k,\ell}^1(u_{r_{ij}})$ and $\chi_{k,\ell}^2(u_{r_{ij}})$) is denoted by Λ_ℓ, and the amount of k-expansion waves between these characteristics at time t, $T_0 \leq t \leq T_1$, is denoted by $w_{k,\ell}^+(t)$. From the results in Section 9 it follows that

$$(11.2) \quad \sum_{\ell=1}^{N} w_{k,\ell}^+(t) \geq \sum_{\ell=1}^{N} w_k^+(T_1;u_{r_{ij}})\Big|_{I_\ell(u_{r_{ij}})} - \sum_{\ell=1}^{N} P_{\lambda_i}(\Lambda_\ell;u_{r_{ij}})$$

$$+ O(1) \sum_{\ell=1}^{N} Q(\Lambda_\ell;u_{r_{ij}})$$

and that $\overline{\chi}_{k,\ell}^1 \equiv \overline{\chi}_{k,\ell}^1(u_{r_{ij}})$ and $\overline{\chi}_{k,\ell}^2 \equiv \overline{\chi}_{k,\ell}^2(u_{r_{ij}})$ are Lipschitz curves propagating with k-characteristic speed λ_k. Since $\chi_{k,\ell}^1(u_{r_{ij}})$ always lies to the left of $\chi_{k,\ell}^2(u_{r_{ij}})$, we may define $\overline{\chi}_{k,\ell}^1$ and $\overline{\chi}_{k,\ell}^2$ so that

§14. Points of interactions

THEOREM 14.1. Suppose that $Q + C$ has nonzero pointed measure at (x_0, t_0) for the weak solution $u_0(x,t)$. Then $u_- \equiv u(x_0-0, t_0) \neq u_+ \equiv u(x_0+0, t_0)$ and the outgoing waves in $u_0(x, t_0)$ at (x_0, t_0) are close to the solution of the Riemann problem (u_-, u_+) and the incoming waves are compression waves in the following sense. There exist Lipschitz continuous curves L_i^+, L_i^-, $i = 1, 2, \ldots, n$, through (x_0, t_0) such that the total variation of $u_0(x, t)$ outside $\Omega_i^{\pm}$ is small near (x_0, t_0), Figure 14.1. Let (u_{i-1}, u_i) be the i-waves, $i = 1, 2, \ldots, n$, in the solution of the Riemann problem (u_-, u_+), and (u_{i-1}, u_i) consists of i-discontinuities $(u_{i,j-1}, u_{i,j})$, j odd, and i-rarefaction waves $(u_{i,j-1}, u_{i,j})$, j even. $u_0(x,t)|_{G_i^+}$, $i = 1, 2, \ldots, n-1$, $u_0(x,t)|_{G_0}$ and $u_0(x,t)|_{G_n}$ tend to u_i, u_- and u_+,

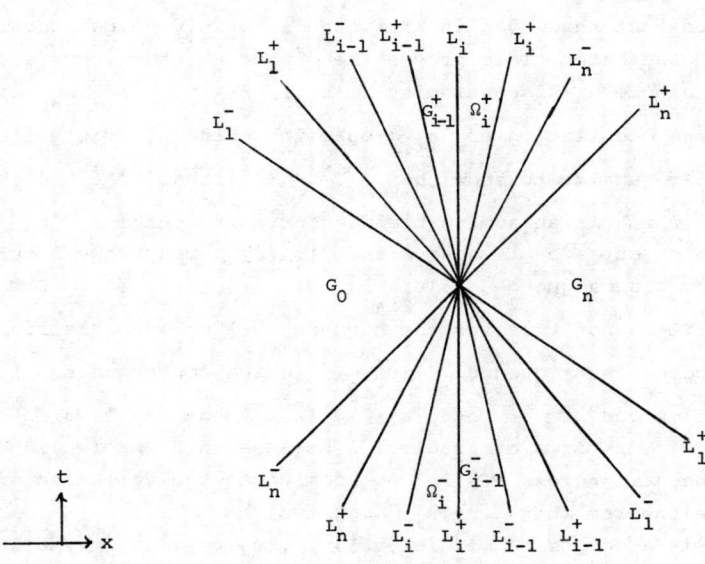

Figure 14.1

respectively, as (x,t) tends to (x_0, t_0). When $(u_{i,j-1}, u_{i,j})$, j odd, is a simple (c.f., Section 13) i-discontinuity, then there exists a Lipschitz continuous curve $\Gamma_{i,j}$ issued at (x_0, t_0) in Ω_i^+ such that $\Gamma_{i,j}$ propagates with speed $\sigma(u_{i,j-1}, u_{i,j})$ at (x_0, t_0). When $(u_{i,j-1}, u_{i,j})$, j odd, is a composite of several i-discontinuities, then we may define a set $\{\Gamma_{i,j}\}$ which consists of several Lipschitz continuous curves issued at (x_0, t_0) in Ω_i^+ such that each propagates with the speed $\sigma(u_{i,j-1}, u_{i,j})$ (c.f., Theorem 13.1). Between $\Gamma_{i,j}$ and $\Gamma_{i,j+2}$,

TABLE OF CONTENTS

1. Introduction   1
2. Jump relation.   6
3. Admissibility criterion. 12
4. Resolution of discontinuities. 15
5. Interaction of elementary waves, I 18
6. Stability of wave pattern. 24
7. Interactions of elementary waves II. 35
8. Nonlinear functional 42
9. Wave partition 46
10. Convergence of approximate solutions 50
11. Expansion waves. 54
12. Continuity points. 60
13. Curves of discontinuity. 62
14. Points of interactions 68
15. Regularity of the solution 73
16. Asymptotic behavior of the solution. 74
17. Linear and nonlinear waves 75

Abstract

We consider a system of n conservation laws:

$$\frac{\partial u}{\partial t} + \frac{\partial f(u)}{\partial x} = 0.$$

The system is assumed to be strictly hyperbolic, but not necessarily genuinely nonlinear in the sense of [19]. Our purpose is to study the regularity, large-time behavior and the approximation of the solution of the initial-value problem. Our analysis is based on the random choice method, [15], [31], using the solution of Riemann problem, [23], as building blocks.

The solution tends to elementary waves determined by the values of the initial data at $x = \pm\infty$. The solution is continuous outside a countable Lipschitz continuous curves of jump discontinuities and is admissible in the sense of [22]. At a point of interaction the incoming waves are compression waves and the outgoing waves are elementary waves determined by the right and left limits of the solution at the point. For each curve of jump discontinuity in the exact solution, there exists a corresponding approximate discontinuity in the approximate solution provided that the mesh length is not large compared to the strength of the discontinuity. The convergence of approximate solutions is locally uniform at points of continuity. In particular we show that the random choice method yields sharp admissible discontinuities, a property not shared by usual finite difference schemes.

The proof of these results is based on the estimates on the local change of wave patterns due to nonlinear interaction. The basic technique in studying the evolution of wave patterns is the   partition method initiated in [31]. The decay theory of [16] is not suitable here because the system is not genuinely nonlinear.

ADMISSIBLE SOLUTIONS

OF HYPERBOLIC CONSERVATION LAWS

§1. Introduction

Consider the initial value problem for conservation laws:

(1.1)
$$\frac{\partial u}{\partial t} + \frac{\partial f(u)}{\partial x} = 0, \qquad t \geq 0, \qquad -\infty < x < \infty,$$

(1.2)
$$u(x,0) = u_0(x), \qquad -\infty < x < \infty,$$

where u is an n-vector, f(u) is an n-vector-valued smooth function of
u, n any positive integer, and the $u_0(x)$ is a given bounded function.
We assume that system (1.1) is strictly hyperbolic, that is, $\partial f(u)/\partial u$ has
real and distinct eigenvalues $\lambda_1(u) < \lambda_2(u) < \cdots < \lambda_n(u)$ with right eigen-
vectors $r_i(u)$: $(\partial f(u)/\partial u) r_i(u) = \lambda_i(u) r_i(u)$. It is well known that a quasilinear system in general does not have global smooth solutions, [20], [18],
[32]. For this reason, we study the weak solutions containing discontinuities. A bounded measurable function u(x,t) is a weak solution of (1.1)
and (1.2) if

(1.3)
$$\iint_{t \geq 0} [u \frac{\partial \phi}{\partial t} + f(u) \frac{\partial \phi}{\partial x}] dx dt + \int_{t=0} u_0 \phi dx = 0$$

for any smooth function ϕ with compact support in $t \geq 0$. One of the
main features of the theory for quasilinear hyperbolic systems is that weak
solutions for a given initial value problem are in general not unique. One
needs an admissibility criterion to select physically relevant solutions.
Such a criterion is usually called entropy condition, as it was motivated
by the second law of thermodynamics for gas dynamics. Thus when a weak
solution is constructed, it is important to make sure that it is admissible.
It is also interesting to study the effect of the admissibility criterion
and the nonlinearity of the system on the bahavior of the solution.

Our interest in characteristics fields which are not genuinely nonlinear or linear degenerate is partly motivated by physical considerations.

Received by the editor July 16th, 1980.

Partially supported by NSF Grant and an Alfred Sloan Foundation Fellowship.

Although equations in fluid dynamics, such as gas dynamics equations, magnetohydrodynamics equations and shallow water wave equations, have genuinely nonlinear or linear degenerate fields, fluid with chemicals, [8],
and multi-phase flows are modeled by equations with nonlinearity weaker
than the genuinely nonlinearity. This is so, roughly speaking, because a
linear combination of convex functions may not be convex. Moreover, equations in solid mechanics cannot be genuinely nonlinear, because the stress-
strain relations for shear waves, torsion waves, etc. are odd relations
and an odd function cannot be genuinely nonlinear, [1], [10].

A stability criterion for genuinely nonlinear system was proposed by
Lax [19] where he solved the Riemann problem (1.1) with data

$$(1.4) \qquad u(x,0) = \begin{cases} u_\ell & \text{for } x < 0, \\ u_r & \text{for } x > 0, \end{cases}$$

where u_ℓ and u_r are two constant states. The solution consists of
rarefaction waves and shock waves. The stability criterion is imposed on
the shock waves and requires that the characteristic curves of certain
family impinge on the shock curve from both sides in the forward time
direction. In [15], Glimm introduced the celebrated random choice method
and solved the general initial value problem. The scheme involves a random sequence and approximates general solutions by solutions of Riemann
problem. The deterministic version of the scheme was obtained in Liu [31].
The decay theory for solutions of system of two equations with periodic or
compact initial data was developed in the difficult paper of Glimm and Lax
[16]. The theory was applied by DiPerna [14] to study the regularity of
the solution. The asymptotic behavior of the general solutions of n-
conservation laws, $n \geq 2$, with linear degenerate and genuinely nonlinear
fields such as equations of gas dynamics and magnetohydrodynamics was obtained by Liu [27], [29], [30]. These works show that the solution of
(1.1), (1.2) tends to the solution of (1.1), (1.4) with $u_\ell = u_0(-\infty)$,
$u_r = u_0(+\infty)$. The L_1 convergence of the solution with compact initial data
to N-waves was proved by DiPerna [13] for two conservation laws based on
the decay result of Glimm and Lax [16]. The L_1 convergence to N-waves
and traveling waves of solutions for general system was obtained in Liu
[28], [30]. The basic mechanism for the striking large-time behavior of
the solution is that shock waves compress and combine due to the stability
criterion and rarefaction waves expand and cancel with shock waves due to
the genuine nonlinearity of the system. Waves pertaining to the linear
degenerate fields behave linearly and tend to traveling waves. Except for
the polytropic gas dynamics equations [33], [2], [34], [12], [26] and two
conservation laws, [16], the existence of the solution has been exhibited
only when the initial data have small total variation [15].

An admissibility criterion for general conservation laws, not

necessarily genuinely nonlinear, was proposed in Liu [22], [23]. Subject
to this criterion, the Riemann problem is uniquely solved in the class of
centered rarefaction waves and discontinuity waves. In contrast to genuinely nonlinear systems, elementary waves also contain one-sided and two-
sided contact discontinuities. These are described in Sections 2, 3, and
4. The composite wave pattern of these elementary waves is less stable
than the shock waves and rarefaction waves for genuinely nonlinear systems.
For instance, when a discontinuity is close to a composite of several
weaker discontinuities, a small perturbation of the discontinuity may split
the discontinuity into several discontinuities. Nevertheless, in Section
6 we will show that a wave pattern is related to its boundary states stably
in a general sense. When a wave pattern is noninteracting, such as those in
a solution of the Riemann problem, it depends smoothly on its boundary
states (Lemma 6.3). When a wave pattern is weakly interacting, it is close
to a wave pattern in a solution of the Riemann problem (Lemma 6.4). This
may be viewed as the stability property of the admissibility criterion.

When two adjacent waves of the same family make a zero angle, they
can be linearly superimposed to form a new wave of that family. In Section 5 we study the interaction of waves of the same family where we observe
that the effect of nonlinear interaction is proportional to the angle between the waves before the interaction. It is also observed that the combining of two waves of the same family and the same direction may cause
the decrease in the total amount of expansion waves. On the other hand,
the cancelling of two waves of the opposite direction may cause the increase in the total amount of expansion waves. These two effects are not
present for genuinely nonlinear or linearly degenerate fields and can be
understood easily for scalar conservation law. The results obtained in
Sections 5 and 6 are used to study the complete local interaction of elementary waves in Section 7. It is shown that the interaction of waves of
different families causes a change of the wave strength and the wave pattern of second order of the strength of incoming waves.

The global estimates of nonlinear interactions are obtained in Section 8 where we introduce a nonlinear functional based on the local estimates in the previous section. The major difference between our functional
and those of [15] and [30] is that the amount of the potential interaction
between waves of the same family is proportional to the angle between the
waves and the product of the strength of the interacting waves. This is
natural since no interaction is expected when the angle is zero. It is
also necessary because when a characteristic field is not genuinely nonlinear, waves of finite strength may make an arbitrary small, yet positive,
angle. Another major difference from previous works is the following: To
measure the strength of waves, one may use the Riemann invariants when
$n = 2$, [15], [33], [27], or the characteristic speed when the system is
genuinely nonlinear, [30], or any Euclidean distance, [15]. Although
these strengths are equivalent to one another, each choice has its own advantage in studying certain aspects of the wave interaction. For our

purpose we choose one of the component of the vector u to measure the
strength of waves. This has an advantage in studying the evolution of the
speed and the strength of discontinuities during the interaction (c.f.
(iii) of Lemma 5.1).

The basic technique in studying the global evolution of wave patterns
is the wave partition method described in Section 9. The method was introduced in [31] to study the evolution of wave speed and strength. For the
purpose of studying the convergence to the weak solution this is sufficient. Also, for genuinely nonlinear system, wave strength is directly
related to the wave type. For the general system, however, we need a
refined partition based on the results on the stability of waves of Section 6 to study the evolution of wave types. This allows us to define
characteristic curves traveling with characteristic speed at all time or
discontinuity speed at all time. The purpose of introducing characteristics of these types is to investigate the conservation of wave types between characteristics and thus allow us to study the expansion of rarefaction waves in Section 11 and also the compression waves in the subsequent
sections.

In Section 10 we study the pointwise convergence of approximate solutions using the elementary real analytic method and the global estimates
on wave interactions in Sections 7 and 8. This implies the L_1 convergence of [15]. A stronger local convergence theorem is proved in Section
12 using the result on the expansion waves of the previous section. This
follows from the observation that when the exact solution is continuous at
a point, the approximate solutions have small local oscillation, though
may not have small local variation. When an approximate solution has a
small amount of rarefaction waves near a point, then the local oscillation
is also small. (This is clear for genuinely nonlinear systems, for general
systems the strength of a wave is not proportional to the jump of the characteristic speed across the wave and the result is a consequence of the
stability of wave patterns due to the admissibility criterion.) The solution cannot possess a large amount of local rarefaction waves at a point
of continuity because of the expansion of these waves as studied in Section 11.

When the solution is discontinuous at a point and there is no pointed
measure of wave interaction at the point, there exists a Lipschitz continuous curve of admissible jump discontinuity through the point, Section 13.
In fact, it is shown that there already exists a corresponding discontinuity in the approximate solution, provided that, of course, the mesh
length is not large compared to the strength of the discontinuity. This
is done in three steps: First we use the arguments in the previous section to show that the amount of expansion waves near the point is small.
Then through an elaborate argument we show that there exist at most two
strong compression waves which are weakly interacting. If this were not
the case the compression waves would combine and create strong measure of

interaction. In other words, the wave pattern near the point is weakly
interacting. Finally we apply the stability result of Section 6 to conclude that the local behavior of the solution resembles the solution of a
Riemann problem which in this case is a discontinuity. Since the discontinuities in the approximate solutions are admissible by our choice, it
follows that when the random sequence is equidistributed the discontinuity
in the exact solution is also admissible and Lipschitz continuous in a
neighborhood of the point. When the discontinuity is a composite of
weaker discontinuities at the point, the discontinuity curve may split
into weaker discontinuity curves in a small neighborhood of the point.
The wave partition method is strong enough to handle this situation because
a single discontinuity can be partitioned and several different generalized
characteristics can be defined along the discontinuity and these characteristics may not coincide at latter or earlier time. The theory of
Glimm-Lax [16] (c.f. [14]) depends essentially on the fact that characteristics curves near a shock curve must impinge toward the shock curve in
the forward time direction at a positive angle and only one generalized
characteristic curve lies on a shock wave, and so it is not suitable for
the present study.

The admissibility of solutions obtained by random choice method was
verified by Lax [21] for genuinely nonlinear systems endowed with entropy
and entropy flux functions. This method is not applicable even for general two conservation laws which is not genuinely nonlinear, [7]. The
regularity result for genuinely nonlinear two conservation laws, [15],
[13], also implies the admissibility of the solution. Our result on the
sharpness of the discontinuities in the approximation solutions justifies
the applications of the random choice method to physical situations which
demand a sharp resolution of discontinuities, [3], [4].

The local behavior of the solution at a point of interaction is
studied in Section 14. We combine the arguments of previous two sections
to show that the incoming waves are compression waves and outgoing waves
are the solution of the Riemann problem whose data are determined by the
right and left limit of the solution at the point. In Section 15 we combine the results in the previous three sections to study the global regularity of the solution. Because the solution is of bounded variation,
there exist at most countable curves of jump discontinuity.

In Section 16 we show that the asymptotic state of the solution consists of the elementary waves in the solution of the Riemann problem with
data $u_\ell = u_0(-\infty)$ and $u_r = u_0(+\infty)$. This is proved by the similar arguments employed in the study of the regularity of the solution in previous
sections based on the following two facts: The limiting values of the
solution at $x = \pm\infty$ are invariant of time and the total amount of wave
interaction and cancellation after large time is small. The decay of
periodic solution for scalar conservation laws has been obtained by
Dafermos [9] using a different method and of a special solution for two

conservation laws by Greenberg [17]. It would be interesting to investigate the rates for our asymptotic results when the initial data equal the
Riemann data outside a bounded interval.

 So far we are interested in the behavior of nonlinear waves. In Section 17 we note that in the presense of linear waves, the behavior of nonlinear waves is still nonlinear in the sense described in previous sections. As an illustration we consider the gas dynamics equations whose
second characteristic field is always linearly degenerate. The first and
third fields are nonlinear. Since the velocity and the pressure of the gas
are constant across 2-waves, their behavior is nonlinear in the above
sense.

 For a uniqueness theorem for a system not genuinely nonlinear, see
[25]. The uniqueness problem remains open even for genuinely nonlinear two
conservation laws, [35], [11].

 In conclusion we describe our results briefly as follows: The random
choice method approximates discontinuities sharply and yields admissible
and, in a rough sense, piecewise continuous solutions. Moreover, the study
of the Riemann problem is sufficient in understanding the local and large-
time behavior of the solution. These results are new even for scalar conservation law for which we need only to assume the initial data to be of
bounded variation and the second derivative of the flux function f has
isolated zeros. To have an easier reading we advise the reader to inter-
pret our techniques for scalar conservation law. The estimates for scalar
conservation laws involve mainly linear terms.

§2. Jump relation

 Suppose that a weak solution $u(x,t)$ is discontinuous of the first
kind along $x = x(t)$, then it follows from (1.3) that it satisfies the
following jump (Rankine-Hugoniot) condition:

$$(R-H) \qquad x'(t)(u_+ - u_-) \;=\; f(u_+) - f(u_-)$$

$$u_\pm \;\equiv\; u(x(t) \pm 0, t).$$

 To study the discontinuities in the solution we define the R-H
curves $S(u_0)$ through a given state u_0 as follows:

$$s(u_0) \;\equiv\; \{u: \sigma(u - u_0) = f(u) - f(u_0) \text{ for some scalar } \sigma = \sigma(u_0, u)\}.$$

 With this notion the jump condition becomes

$$(R-H) \qquad u_+ \in S(u_-), \qquad x'(t) = \sigma(u_-, u_+).$$

The following two lemmas on R-H curves were obtained in [19] (c.f. [5], [6]). For the sake of completeness, we present a simple proof of these two lemmas. For future use, we define the i-rarefaction curve $R_i(u_0)$, $i = 1,2,\ldots,n,$ to be the integral curve of the vector field $r_i(u)$ through the state u_0.

LEMMA 2.1. In a small neighborhood of u_0, the set $S(u_0)$ consists of n smooth curves $S_1(u_0)$, $S_2(u_0),\ldots,S_n(u_0)$ through the state u_0 so that $S_i(u_0)$ is tangent to $R_i(u_0)$ at u_0 and the speed of discontinuity $\sigma(u_0,u)$ tends to the characteristic speed $\lambda_i(u_0)$ as $u \in S_i(u_0)$ tends to u_0.

PROOF: Write the jump condition (R-H) $\sigma(u-u_0) = f(u) - f(u_0)$ as

$$(G(u)-\sigma)(u-u_0) \;=\; 0$$

$$G(u) \;\equiv\; \int_0^1 \frac{\partial f}{\partial u}(u_0+z(u-u_0))\,dz.$$

Clearly $G(u)$ tends to $\frac{\partial f}{\partial u}(u_0)$ as u tends to u_0. Since (1.1) is strictly hyperbolic we see that for u close to u_0, $G(u)$ has real and distinct eigenvalues. The above identity shows that σ is an eigenvalue and $u - u_0$ a right eigenvector of $G(u)$. Thus σ is close to $\lambda_i(u_0)$ for some $i \in \{1,2,\ldots,n\}$ and the identity is equivalent to

$$\ell_j(u) \cdot (u-u_0) \;=\; 0 \qquad j = 1,2,\ldots,i-1,i+1,\ldots,n.$$

where $\ell_j(u)$ is a j-th left eigenvector of $G(u)$. By the implicit function theorem the above system can be solved to yield a smooth one-parameter family of solutions. This proves the lemma. Q.E.D.

LEMMA 2.2. Suppose that the right eigenvector $r_i(u)$ of $\partial f(u)/\partial u$ is normalized $\|r_i\| = 1$ and μ_i is the arc length along $S_i(u_0)$. Then for u on $S_i(u_0)$

(i) $\dfrac{du}{d\mu_i} = r_i(u_0),$ $\sigma(u_0,u) = \lambda_i(u_0)$ at $u = u_0$

(ii) $\dfrac{d\sigma(u_0,u)}{d\mu_i} = \dfrac{1}{2} r_i \cdot \nabla\lambda_i$ at $u = u_0$

(iii) $\dfrac{d^2u}{d\mu_1^2} = r_i \cdot \nabla r_i$ at $u = u_0.$

PROOF: We have, by differentiating the jump condition (R-H),

(2.5) $\alpha_i \beta_i > 0$ for $u \neq u_0$.

(2.6) $\left| \dfrac{\alpha_i}{u-u_0} \right|$ and β_i are positive and bounded.

and from (iii) of Lemma 2.2 we have

(2.7) $\dfrac{|\alpha_j|}{|u-u_0|^2}$ is bounded for $j \neq i$.

We have from (2.1) $\sim$ (2.3) that

$(2.8)_j$ $\alpha_j \dfrac{d\sigma}{d\mu_i} = (\lambda_j(u)-\sigma)\beta_j,$ $j = 1,2,\ldots,n.$

Thus (i) of the lemma follows from (2.5) and $(2.8)_i$. Since σ is close to λ_i and $\lambda_i \neq \lambda_j$ for $i \neq j$, we see that $\lambda_j(u) - \sigma \neq 0$ and so (ii) follows from (2.8) and (i).   Q.E.D.

LEMMA 2.4. Suppose that u_1 is on $S_i(u_0)$ and in a small neighborhood of u_0 and $|\nabla\lambda_i \cdot r_i(u)| \leq C_0$ for any $u \in S_i(u_0)$ between u_0 and u_1. Then for some constant C_1 depending only on (1.1) there exists $u_2 \in R_i(u_0)$ such that

  (i) $|\sigma(u_0,u_1) - \lambda_i(u_1)| \leq C_0 C_1 |u_0-u_1|,$

  (ii) $|u_2-u_1| \leq C_0 C_1 |u_0-u_1|^2.$

When $\nabla\lambda_i \cdot r_i(u) \geq C_0 > 0$ (or $\nabla\lambda_i \cdot r_i(u) \leq -C_0 < 0$) for all $u \in S_i(u_0)$ between u_0 and u_1, then for some positive constant C_2 we have

  (iii) $\lambda_i(u_1) - \sigma(u_0,u_1) \geq C_2 C_0 |u_0-u_1|$

                           (or $\lambda_i(u_1) - \sigma(u_0-u_1) \leq -C_2 C_0 |u_0-u_1|$);

  (iv) $\left| \dfrac{d\sigma}{d\mu_i} \right| \geq C_2 C_0.$

   PROOF: We have from (2.6), (2.7) and $(2.8)_j$, $j \neq i$, that

(2.9) $\beta_j = \dfrac{\alpha_j}{\alpha_i} \dfrac{\lambda_i - \sigma}{\lambda_j - \sigma} = O(1) |u-u_0|(\lambda_i - \sigma),$ $j \neq i.$

It follows from $(2.8)_i$ that

            $\alpha_i \dfrac{d(\lambda_i - \sigma)}{d\mu_i} = \alpha_i \dfrac{d\lambda_i}{d\mu_i} - (\lambda_i - \sigma)\beta_i$

and from (2.4) that

$$\frac{d\lambda_i}{d\mu_i} = \sum_{j=1}^{n} \beta_j (\nabla\lambda_i \cdot r_j).$$

The above two equalities and (2.8) yield

(2.10) $\alpha_i \dfrac{d(\lambda_i-\sigma)}{d\mu_i} = \beta_i[\alpha_i\nabla\lambda_i\cdot r_i-(\lambda_i-\sigma)] + O(1)|u-u_0|(\lambda_i-\sigma).$

Note that (2.10) holds for any nonsingular parameter μ_i along $S_i(u_0)$. In view of (2.6) we may choose, for our convenience, $\mu_i = \alpha_i/\beta_i$ and (2.10) becomes

(2.11)
$$\frac{d\phi}{d\mu_i} + O(1)\phi = \alpha_i\nabla\lambda_i \cdot r_i$$

$$\phi \equiv \mu_i(\lambda_i-\sigma), \qquad \mu_i = \alpha_i/\beta_i$$

which can be solved to obtain

(2.12) $\mu_i(\lambda_i-\sigma) = \exp(O(1)) \cdot \displaystyle\int_0^{\mu_i} \mu_i(\nabla\lambda_i\cdot r_i)\exp(O(1))d\mu_i$

(i) and (iii) of the lemma follows easily from the above estimate. Finally (ii) follows from (i) and (2.9), and (iv) is a consequence of (iii) and (2.8).   Q.E.D.

As an easy consequence of the above lemma, we see that when an i-field is linearly degenerate, i.e., $\nabla\lambda_i\cdot r_i \equiv 0$, then $R_i(u_0) = S_i(u_0)$ and $\sigma(u_0,u) = \lambda_i(u_0) = \lambda_i(u)$ for any u on $S_i(u_0)$. In other words, an i-wave is a two-sided contact discontinuity.

LEMMA 2.5. Suppose that u_0 and $\tilde{u}$ are close and

(2.13) $\tilde{\sigma}(u_0-\tilde{u}) = f(u_0) - f(\tilde{u}) + K$

for some scalar $\tilde{\sigma}$ and a vector K satisfying

$$\|K\| = \varepsilon\|u_0 - \tilde{u}\|$$

for some small ε. Then there exists $\tilde{u}_2$ on $S(u_1)$ such that
   (i) $\|u-\tilde{u}\| = O(1)\|K\|$,
   (ii) $|\tilde{\sigma} - \sigma(u_0,u)| = O(1)\varepsilon.$

PROOF: It is clear from the implicit function theorem that $\tilde{u}$ is close to $R_i(u_0)$ for some $i \in \{1,2,\ldots,n\}$ and $\tilde{\sigma}$ is close to $\lambda_i(u_0)$. Choose any nonsingular parameter μ_i along $R_i(u_0)$ and $S_i(u_0)$ and let u be the state on $S_i(u_0)$ with $\mu_i(u) = \mu_i(\tilde{u})$:

(2.14) $\sigma(u-u_0) = f(u) - f(u_0),   \qquad \sigma \equiv \sigma(u_0,u)$.

We have from (2.13) and (2.14) that

(2.15) $\tilde{\sigma}(\tilde{u}-u) - f(\tilde{u}) - f(u) = (\sigma-\tilde{\sigma})(u-u_0) + K$.

Write $\tilde{u} - u = \sum_{j=1}^{n} a_j r_j(u)$. By our choice of i and u we have

(2.16) $a_{j_0} \geq |a_j|, \quad j = 1,2,\ldots,n, \quad \dfrac{a_{j_0}}{\|u-\tilde{u}\|}$ and $\dfrac{\|u-\tilde{u}\|}{a_{j_0}}$ both bounded

for some $j_0 \neq i$. Thus we have from (2.15) that

(2.17) $\displaystyle\sum_{j=1}^{n} a_j(\tilde{\sigma}-\lambda_j(u))r_j(u) = (\sigma-\tilde{\sigma})(u-u_0) + K + O(1)\|u-\tilde{u}\|(u-\tilde{u})$.

Since $\tilde{\sigma}$ is close to $\lambda_i(u)$ and $\lambda_i \neq \lambda_j$ we have from (2.16) and (2.17) that

(2.18) $\|u-\tilde{u}\| = O(1)[|\sigma-\tilde{\sigma}|\,\|u-u_0\| + K]$.

On the other hand, by comparing the $r_i(u)$ component of both sides of (2.17) we have

(2.19) $|\sigma-\tilde{\sigma}|\,\|u-u_0\| = O(1)[K + a_i|\tilde{\sigma}-\lambda_i(u)|]$.

Since $|\tilde{\sigma}-\lambda_i(u)|$ is small and $|a_i| = O(1)\|u-\tilde{u}\|$, we have from (2.18) and (2.19) that (i) of the lemma holds. (ii) follows from (i) and (2.19).
                                                                Q.E.D.

Remark 2.6. The above lemmas hold for u in a neighborhood N of the initial state u_0. It is clear from our estimates that the size of N is inverse proportional to the maximum norm of the second derivatives of the flux function f(u) of (1.1). Examples can be constructed to show that $S_i(u_0)$ may join $S_j(u_0)$ for $i \neq j$. Thus the neighborhood N cannot be arbitrary large. However, for some physical models such as gas dynamics equations, the R-H curves are well behaved.

§3. Admissibility criterion

Definition 3.1. A discontinuity (u_-,u_+) is an admissible i-discontinuity if $u_+ \in S_i(u_-)$, $i \in \{1,2,\ldots,n\}$ and it satisfies the following entropy condition

$$(E) \qquad \sigma(u_-,u_+) \leq \sigma(u_-,u)$$

for any $u \in S_i(u_-)$ between u_- and u_+.

It is not difficult to show that the above definition of condition (E) is equivalent to the following

$$(E) \qquad \sigma(u_-,u_+) \geq \sigma(u_+,u)$$

for any $u \in S_i(u_+)$ between u_- and u_+.

We now derive some basic properties for admissible discontinuities, [23].

LEMMA 3.2. Suppose that (u_-,u_+) is an admissible i-discontinuity and u_+ is close to u_-. Then the following stability condition holds:

$$\lambda_i(u_-) \geq \sigma(u_-,u_+) \geq \lambda_i(u_+) .$$

PROOF: The first inequality follows immediately from condition (E) since $\lim_{u \to u_-} \sigma(u_-,u) = \lambda_i(u_-)$ by (i), Lemma 2.2. From condition (E), $\sigma(u_-,u)$ is nonincreasing at $u = u_+$ as u moves from u_- toward u_+. Thus the second inequality follows from (i), Lemma 2.3.   Q.E.D.

The following lemma is an easy consequence of (iv) of Lemma 2.3.

LEMMA 3.3. When an i-field is genuinely nonlinear; that is, $\nabla\lambda_i \cdot r_i \neq 0$, then condition (E) for (u_-,u_+) is reduced to the following shock stability condition, [21]:

$$(L) \qquad \lambda_i(u_-) > \sigma(u_-,u_+) > \lambda_i(u_+) .$$

In this case (u_-,u_+) is called an i-shock wave.

When an i-field is linearly degenerate, that is, $\nabla\lambda_i \cdot r_i \equiv 0$, then condition (E) is always satisfied and (u_-,u_+) is called a two-sided contact discontinuity because $\sigma(u_-,u_+) = \lambda_i(u_-) = \lambda_i(u_+)$.

Except in Section 17, from here on we will be concerned with the behavior of nonlinear waves. In other words, we assume that each characteristic field is not linear degenerate in any region. This is made precise as follows: We set

$$\Omega_\varepsilon^i \equiv \{u: \ |\nabla\lambda_i \cdot r_i(u)| \leq \varepsilon\}, \qquad \varepsilon > 0.$$

We assume that in each bounded set in u space

(3.1)
$$\Omega_0^i = \bigcup_{k=1}^{K} \Omega_0^{i,k}$$

where k is a __finite__ number and each $\Omega_0^{i,k}$ is a m-dimensional manifold, $m \leq n-1$, which is __transversal__ to the vector field $r_i(u)$. This implies that along each R_i curve λ_i has finite number of critical points. Moreover, if we denote by $\phi(\varepsilon, R_i)$ the length of subintervals of $R_i \cap \Omega_\varepsilon^i$, then

$$\phi(\varepsilon, R_i) \longrightarrow 0 \quad \text{as} \quad \varepsilon \longrightarrow 0_+ .$$

We know from Lemma 2.4 that either when $|\nabla\lambda_i \cdot r_i|$ is small or waves are weak $\phi(\varepsilon, S_i)$ can be defined by replacing R_i curves by S_i curves and satisfies similar estimate. We set

$$\phi_i(\varepsilon) \equiv \max\{\phi(\varepsilon, R_i) + \phi(\varepsilon, S_i); \ R_i \text{ and } S_i \text{ any refraction}$$
$$\text{and R-H curves under consideration}\}.$$

Our nonlinearity hypothesis (3.1) thus becomes

(3.2)
$$\phi_i(\varepsilon) \longrightarrow 0 \quad \text{as} \quad \varepsilon \longrightarrow 0_+, \qquad i = 1,2,\dots,n.$$

__LEMMA__ 3.3. Suppose that (3.2) holds. Then along $S_i(u_0)$ the following holds:

   (i) $\sigma(u_0,u)$ has finite number of critical points, and if $\sigma(u_0,u)$ attains a local minimum (maximum) at u then

   (ii) in a small neighborhood of u, $\nabla\lambda_i \cdot r_i > 0$ ($\nabla\lambda_i \cdot r_i < 0$) along $R_i(u) - \{u\}$

   (iii) $\sigma(u_0,u)$ have an inflection point and is nonincreasing (nondecreasing) at u only if $\nabla\lambda_i \cdot r_i = 0$ at u, and moreover, in a small neighborhood of u, $\nabla\lambda_i \cdot r_i > 0$ ($\nabla\lambda_i \cdot r_i < 0$) between u and u_0 and $\nabla\lambda_i \cdot r_i < 0$ ($\nabla\lambda_i \cdot r_i > 0$) along $R_i(u)$ not between u and u_0.

Here for definiteness we have choosen r_i to be in the same direction as $u - u_0$.

   PROOF: We will only prove that if $\sigma(u_0,u)$ has a critical point at u and $\nabla\lambda_i \cdot r_i \neq 0$ at u, then $\sigma(u_0,u)$ attains a local minimum if and only if $\nabla \lambda_i \cdot r_i > 0$ ($\nabla \lambda_i \cdot r_i < 0$) at u. The rest of the lemma can be proved by similar arguments. For this we have to show that if $\dfrac{d\sigma}{d\mu_i} = 0$ and $\nabla \lambda_i \cdot r_i \neq 0$ at u then

$$(3.3) \qquad \frac{d^2\sigma}{d\mu_i^2} > 0 \quad \text{if and only if} \quad \nabla \lambda_i \cdot r_i > 0$$

at u for any nonsingular parameter μ_i along $S_i(u_0)$. Choose μ_i to be arc length along $S_i(u_0)$. We know from Lemma 2.3 that $\dfrac{d\sigma}{d\mu_i} = 0$, $\lambda_i = \sigma$ and $\dfrac{du}{d\mu_i} = r_i$ at u. Thus from (2.2) we have

$$(\frac{\partial f}{\partial u} - \lambda_i)(u'' - r_i') \;=\; \sigma''(u - u_0) - \lambda_i' r_i \, .$$

Since $(\frac{\partial f}{\partial u} - \lambda_i)(u'' - r_i')$ is a linear combination of $\{r_j, \; j \neq i\}$ and $u - u_0$ is in the same direction as r_i, (3.3) follows from the above equality.

Q.E.D.

LEMMA 3.4. Suppose that (u_-, u_+) is an admissible i-discontinuity and $\nabla \lambda_i \cdot r_i \leq -\delta$ at u_+ for some positive number δ, and r_i is in the same direction as $u_+ - u_-$. Then either

(i) $\|u_- - u_+\| \leq \delta$

or there exists $C_1 > 0$ depending only on (1.1) such that

(ii) $\sigma(u_-, u_+) - \lambda_i(u_+) \geq C_1 \delta^2 > 0$.

PROOF: Suppose that $\|u_-, u_+\| \geq \delta$. Since $\nabla \lambda_i \cdot r_i \leq -\delta$ at u_+, by continuity, there exists u_0 on $S_i(u_-)$ between u_- and u_+ such that

$$(3.4) \qquad \|u_+ - u_0\| \geq C_0 \delta \quad \text{for some} \quad C_0 > 0 \quad \text{and}$$

$$(3.5) \qquad \nabla \lambda_i \cdot r_i \leq -\frac{\delta}{2} \quad \text{along} \quad S_i(u_-) \quad \text{between} \quad u_0 \quad \text{and} \quad u_+.$$

We claim that

$$(3.6) \qquad \sigma(u_-, u_0) \geq \lambda_i(u_0).$$

This is so because $\sigma(u_-, u_+) \geq \lambda_i(u_+)$ by condition (E) (c.f. Lemma 3.2) and, again because of condition (E) for (u_-, u_+), $\sigma(u_-, u)$ does not have any critical point between u_0 and u_+ by (3.5) and Lemma 3.3. Finally we integrate (2.11) from u_0 to u_+ (c.f. (2.12)) and use (3.4) $\sim$ (3.6) to conclude that (ii) holds for some positive constant C_1. Q.E.D.

§4. Resolution of discontinuities

Our purpose in this section is to describe a procedure in solving the Riemann problem (1.1) and

$$(4.1) \qquad u(x,0) \;=\; \begin{cases} u_\ell & \text{for } x < 0 \\[2mm] u_r & \text{for } x > 0 \end{cases}$$

where u_ℓ and u_r are two constant states. Since (1.1) and (4.1) are both invariant under the transformation $(x,t) \to (cx,ct)$, $c > 0$, we seek for solutions of the form

$$(4.2) \qquad u(x,t) \;=\; \psi(\tfrac{x}{t}).$$

The function ψ is in general piecewise smooth. Suppose that ψ is smooth for $\xi_0 < \tfrac{x}{t} < \xi_1$, $u_0 \equiv \psi(\xi_0)$, $u_1 \equiv \psi(\xi_1)$. Then it follows easily from (1.1) that for some i, $1 \le i \le n$,

$$(4.3) \qquad \begin{aligned} &\psi(\xi) \in R_i(u_0) \\[2mm] &\lambda_i(\psi(\xi)) \;=\; \xi, \qquad \xi_0 < \xi < \xi_1. \end{aligned}$$

Such a solution is called an (centered) i-rarefaction wave (u_0,u_1). Conversely, when $u_1 \in R_i(u_0)$, $\lambda_i(u_1) > \lambda_i(u_0)$, and $\nabla\lambda_i \cdot r_i \neq 0$ between u_0 and u_1, then (u_0,u_1) forms a centered i-rarefaction wave satisfying (4.2) and (4.3). Another kind of elementary waves is admissible i-discontinuities. If $u_1 \in S_i(u_0)$ and (u_0,u_1) is admissible, then

$$(u_0,u_1)(x,t) \;=\; \begin{cases} u_0 & \text{for } x < \sigma(u_0,u_1)t \\[2mm] u_1 & \text{for } x > \sigma(u_0,u_1)t \end{cases}$$

is called a centered i-discontinuity. We now show that when u_ℓ is close to u_r, the Riemann problem (1.1), (4.1) can be solved in the class of elementary waves described above. Since an i-rarefaction wave propagates with speed λ_i, (4.3), and an admissible i-discontinuity (u_-,u_+) has speed between $\lambda_i(u_-)$ and $\lambda_i(u_+)$, Lemma 3.2, and the system is strictly hyperbolic, the solution (4.2) consists of constant states $u_\ell = u_0$, $u_1,\ldots,u_{n-1}$, $u_n = u_r$ and elementary i-waves relating u_{i-1} and u_i, $i = 1,2,\ldots,n$.

As a first step toward solving the Riemann problem, we construct curves $T_i(u_0)$, $i = 1,2,\ldots,n$, through any given state u_0 with the property that any state $u \in T_i(u_0)$ can be connected to u_0 on the left by centered i-waves. The curve $T_i(u_0)$ is tangent to some R_i or S_i

curve and will be parameterized by μ_i. Since S_i is tangent to R_i at initial points, Lemma 2.1, in a small neighborhood of u_0, we may choose μ_i to be any fixed nonsingular parameter along R_i curves. For definiteness we set $\mu_i(u_0) = 0$ and μ_i increases in the direction of r_i. We now construct $T_i(u_0)$ for $\mu_i \geq 0$. Given $\mu_* > 0$, we will find a (unique) state $u_* \in T_i(u_0)$ with $\mu_i(u_*) = \mu_*$. Let $u_1 \in S_i(u_0)$ be defined as follows:

(4.4) (u_0, u_1) satisfies condition (E), $0 \leq \mu_i(u_1) \leq \mu_*$, and for any $u \in S_i(u_0)$, $\mu_i(u_1) < \mu_i(u) \leq \mu_*$, (u_0, u) does not satisfy condition (E).

It is clear that these properties uniquely determine the state u_1. Moreover if $0 < \mu_i(u_1) < \mu_*$, then $\sigma(u_0, u)$ attains a minimum at $u = u_1$. Thus, in particular, $\sigma(u_0, u_1) = \lambda_i(u_1)$, Lemma 2.3, and (u_0, u_1) is usually called a (right-sided) i-contact discontinuity. When $\mu_i(u_1) = \mu_*$, set $u_* = u_1$ and we are done. When $0 \leq \mu_i(u_1) < \mu_*$, it follows from Lemma 3.3 that $\nabla \lambda_i \cdot r_i \geq 0$ at u_1 and for $u \in R_i(u_1)$, $\mu_i(u) - \mu_i(u_1)$ positive and small, $\nabla \lambda_i \cdot r_i(u) > 0$ and $\dfrac{d\sigma(u_1, u)}{d\mu_i} > 0$. Thus it follows from Lemma 2.2 that $\sigma(u_1, u)$ increases as u moves away from u_1 along $S_i(u_1)$. Consequently, if there exists u on $S_i(u_1)$, $\mu_i(u_1) < \mu_i(u) \leq \mu_*$, such that (u_1, u) satisfies condition (E), then there exists $\tilde{u}$ on $S_i(u_1)$, $\mu_i(u_1) < \mu_i(\tilde{u}) \leq \mu_i(u)$ such that $\sigma(u_1, \tilde{u}) = \lambda_i(u_1)$ and $(u_1, \tilde{u})$ satisfies condition (E). This, along with the property that $\sigma(u_0, u_1) = \lambda_i(u_1)$ and (u_0, u_1) satisfies condition (E), implies that $\tilde{u} \in S_i(u_0)$, $\sigma(u_0, \tilde{u}) = \lambda_i(u_1)$, and $(u_0, \tilde{u})$ satisfies condition (E). (This can be easily proved using the jump condition (R-H) and by contradictions.) This contradicts with the definition of the state u_1. We have thus shown that there does not exist a state u on $S_i(u_1)$, $\mu_i(u_1) < \mu_i(u) \leq \mu_*$, such that (u_1, u) satisfies condition (E). Next we define the state u_2 as follows:

(4.5) $u_2 \in R_i(u_1)$, $\mu_i(u_1) \leq \mu_i(u_2) \leq \mu_*$ and there does not exist $u \in R_i(u_1)$, $\mu_i(u_1) \leq \mu_i(u) < \mu_i(u_2)$, with the property that $(u, \tilde{u})$ satisfies condition (E) for some $\tilde{u} \in S_i(u)$, $\mu_i(\tilde{u}) < \mu_i(u) \leq \mu_*$. Moreover, either $\mu_i(u_2) = \mu_*$ or there exists $u \in S_i(u_2)$, $\mu_i(u_2) < \mu_i(u) \leq \mu_*$ such that (u_2, u) satisfies condition (E).

We have just shown that this state (uniquely) defined above does not equal u_1. If $\mu_i(u_2) = \mu_*$, then the above definition and (iii) of Lemma 2.4 imply that $\nabla \lambda_i \cdot r_i > 0$ along $R_i(u_1)$ between u_1 and u_2. In this case we set $u_* = u_2$ and u_* is related to u_0 by the i-discontinuity

(u_0, u_1) and the i-rarefaction wave (u_1, u_*). If $\mu_i(u_2) < \mu_*$, we define u_3 according (4.4) with u_0 and u_1 replaced by u_2 and u_3 respectively. Definition (4.5) implies that $\mu_i(u_2) < \mu_i(u_3) \leq \mu_*$. As before if $\mu_i(u_3) < \mu_*$, then there does not exist $u \in S_i(u_3)$, $\mu_i(u_3) < \mu_i(u) \leq \mu_*$, such that (u_3, u) satisfies condition (E). In this case, we define the state u_4 by (4.5) with u_1 and u_2 replaced by u_3 and u_4 respectively. Since (u_2, u_3) satisfies condition (E), we have $\sigma(u_2, u_3) \leq \lambda_i(u_2)$. When $\sigma(u_2, u_3) < \lambda_i(u_2)$ then there exists $u \in R_i(u_2)$ between u_2 and u_3 with the property that for some $\tilde{u} \in S_i(u)$, $\mu_i(u) = \mu_i(u_3)$, such that $(u, \tilde{u})$ satisfies condition (E) (c.f. proof of Lemma 6.2 in Section 6) which would contradict the definition of u_3. In other words, $\sigma(u_2, u_3) = \lambda_i(u_2)$ and (u_2, u_3) is usually called a (left-sided) contact discontinuity. When $\mu_i(u_3) < \mu_*$, (u_3, u_4) is an i-rarefaction wave, and $\sigma(u_2, u_3) = \lambda_i(u_2) = \lambda_i(u_3)$. In this case (u_2, u_3) is called a (two-sided) contact discontinuity. It is clear from the above arguments that we may define states $u_0, u_1, u_2, \ldots, u_m = u_*$ with the property that (u_{i-1}, u_i), i odd, are i-discontinuities and (u_{i-1}, u_i), i even, are i-rarefaction waves, Figure 4.1. Of course, it is possible that $m = 1$ and (u_0, u_*) is an i-discontinuity or $u_1 = u_0$, $u_2 = u_*$ and (u_0, u_*) is an i-rarefaction wave. In fact, this is always the case when i-field is genuinely non-linear, [19].

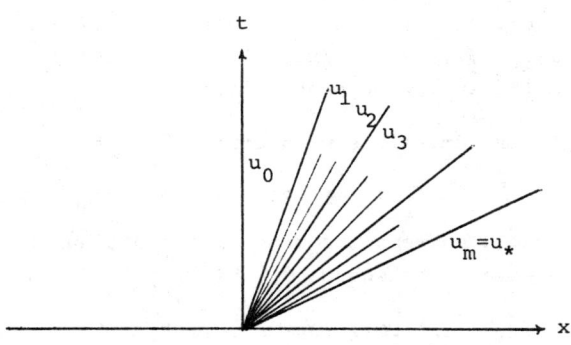

Figure 4.1

The above construction of the curves $T_i(u_0)$ is equivalent to that of [23]. It can be shown that $T_i(u_0)$ is a C^2 curve which is tangent to certain S_i or R_i curve at each point. Since $\{r_1(u), r_2(u), \ldots, r_n(u)\}$ are linear independent because of the strict hyperbolicity of the system (1.1), we may use the implicit functions theorem to find unique states $u_\ell = u_0$, $u_1, \ldots, u_n = u_r$, $u_i \in T_i(u_{i-1})$, $i = 1, 2, \ldots, n$, so that the solution of the Riemann problem (1.1), (4.1) consists of i-waves (u_{i-1}, u_i), $i = 1, 2, \ldots, n$.

We have constructed a solution for the Riemann problem (1.1), (4.1). It turns out that solutions of the form (4.2) are unique [22]. In fact, the solution depends smoothly on the data in a very general sense to be described later in Section 6.

§5. Interaction of elementary waves, I

Our purpose in this section is to study the interaction of an i-discontinuity with either an i-discontinuity or an i-rarefaction wave of the same direction. We will measure the strength of i-waves by a nonsingular parameter μ_i along S_i and R_i. By a linear transformation on the components of the vector $u = (u^i)$ we may assume, without loss of generality, that u^i is strictly monotone along S_i and R_i curves and set $\mu_i \equiv u^i$. Thus the __strength__ of an i-wave (u_0, u_1) is defined to be

$$(5.1) \qquad\qquad | (u_0, u_1) | \;=\; | u_0^i - u_1^i |$$

where u_0^i is the i-th component of u_0, etc.

In the following lemmas, the bound $O(1)$ is proportional to $\max_u |\nabla \lambda_i \cdot r_i(u)|$. We will assume that the waves are weak in the sense that either $\max_u |\nabla \lambda_i \cdot r_i(u)|$ or the strength of these waves is small.

LEMMA 5.1. Suppose that (u_0, u_1) and (u_1, u_2), $u_2^i > u_1^i > u_0^i$, are admissible weak discontinuities with strength $\alpha_1 = u_1^i - u_0^i$ and $\alpha_2 = u_2^i - u_1^i$ and speed $\sigma_1 = \sigma(u_0, u_1)$, $\sigma_2 = \sigma(u_1, u_2)$ respectively. Then there exists $u_* \in S_i(u_0)$ with the following properties:

(i) (u_0, u_*) is admissible with strength $\beta = \alpha_1 + \alpha_2$.

(ii) $\|u_2 - u_*\| = O(1)\beta\theta \min(\alpha_1, \alpha_2)$,

where θ is the angle between incoming i-discontinuities, $\theta \equiv \sigma(u_0, u_1) - \sigma(u_1, u_2) \geq 0$. Moreover, the speed $\sigma \equiv \sigma(u_0, u_*)$ of (u_0, u_*) satisfies:

(iii) $\sigma\beta = \sigma_1\alpha_1 + \sigma_2\alpha_2 + O(1)\beta\theta \min(\alpha_1, \alpha_2)$.

PROOF: Pick $u_* \in S_i(u_0)$ with $u_*^i = u_2^i$ so that (u_0, u_*) has strength

$$(5.2) \qquad\qquad \beta \;=\; \alpha_1 + \alpha_2.$$

From the jump condition (R-H) we have

$$\sigma(u_*^j - u_0^j) \;=\; f^j(u_*) - f^j(u_0),$$

$$(5.3) \qquad \sigma_1(u_2^j - u_1^j) \;=\; f^i(u_2) - f^i(u_1),$$

$$\sigma_2(u_1^j - u_0^j) \;=\; f^j(u_1) - f^j(u_0), \qquad j = 1, 2, \ldots, n,$$

where f^j is the j-component of the vector f. For definiteness, we assume that

(5.4) $\alpha_1 \geq \alpha_2,$ $\alpha_2 = \min(\alpha_1, \alpha_2).$

By (5.2) we may set $\tilde{\sigma}$ by

(5.5) $\tilde{\sigma}\beta \equiv \sigma_1\alpha_1 + \sigma_2\alpha_2,$ $(\tilde{\sigma}-\sigma_1)\alpha_1 + (\tilde{\sigma}-\sigma_2)\alpha_2 \equiv 0.$

We have from (5.2), (5.5) and the jump condition (5.3) for (u_0, u_1) and (u_1, u_2) that for $j = 1, 2, \ldots, n$

(5.6) $\tilde{\sigma}(u_2^j - u_0^j) - [f^j(u_2) - f^j(u_0)]$

$= \tilde{\sigma}(u_2^j - u_0^j) - [\sigma_2(u_2^j - u_1^j) + \sigma_1(u_1^j - u_0^j)]$

$= (\tilde{\sigma}-\sigma_1)(u_2^j - u_1^j) + (\tilde{\sigma}-\sigma_2)(u_1^j - u_0^j)$

$= (\tilde{\sigma}-\sigma_2)(\alpha_1)^{-1}[-(u_2^i - u_1^i)(u_1^j - u_0^j) + (u_2^j - u_1^j)(u_1^i - u_0^i)]$

It follows from Lemma 2.2 that we may choose $\tilde{u}_0$ and $\tilde{u}_2$ on the straight line through u_1 with tangent $r_i(u_1)$ such that

(5.7)
$$\tilde{u}_0^i = u_0,\quad \tilde{u}_2^i = u_2^i,$$
$$\|\tilde{u}_0 - u_0\| = O(1)\alpha_1^2,\qquad \|\tilde{u}_2 - u_2\| = O(1)\alpha_2^2.$$

For $\tilde{u}_0$ and $\tilde{u}_2$ we have

$$-(\tilde{u}_2^i - u_1^i)(u_1^j - \tilde{u}_0^j) + (\tilde{u}_2^j - u_1)(u_1^i - \tilde{u}_0^i) = 0$$

and thus (5.6) and (5.7) yield

(5.8) $\tilde{\sigma}(u_2 - u_0) - [f(u_2) - f(u_0)] = (\tilde{\sigma}-\sigma_2)(\alpha_1)^{-1}[O(1)\alpha_2\alpha_1^2 + O(1)\alpha_2^2\alpha_1]$

$$= O(1)(\tilde{\sigma}-\sigma_2)(\alpha_1\alpha_2 + \alpha_2^2).$$

From (5.2) and (5.4) we have

$$\alpha_1\alpha_2 + \alpha_2^2 \leq \beta \min(\alpha_1, \alpha_2).$$

LEMMA 5.2. Suppose that (u_-,u_1) is an i-rarefaction wave, (u_1,u_+) is an admissible i-discontinuity, and $u_-^i < u_1^i < u_+^i$. Then there exists u_* on $R_i(u_-)$, $u_-^i \le u_*^i \le u_1^i$ and $\tilde{u}_*$ on $S_i(u_*)$, $\tilde{u}_*^i = u_+^i$ with the following properties:

 (i) $\|\tilde{u}_* - u_+\| = O(1)\beta\theta \min(\alpha_1,\alpha_2)$

 (ii) $\sigma\beta = \hat{\lambda}\alpha_1 + \sigma_2\alpha_2 + O(1)\beta\theta \min(\alpha_1,\alpha_2)$

 (iii) $(u_*,\tilde{u}_*)$ is admissible with strength $\beta = \alpha_1 + \alpha_2$.

Here $\alpha_1 \equiv u_1^i - u_*^i \ge 0$ is the strength of <u>canceled</u> rarefaction wave, $\alpha_2 \equiv u_+^i - u_1^i > 0$ is the strength of i-discontinuity before interaction, $\theta = \lambda_i(u_1) - \sigma(u_1,u_+)$ is the angle between these two waves, $\sigma_1 \equiv \sigma(u_*,\tilde{u}_*)$, $\sigma_2 \equiv \sigma(u_1,u_+)$, and $\hat{\lambda}$ is the <u>average</u> <u>speed</u> of the cancelled i-rarefaction wave (u_*,u_1):

$$\hat{\lambda} \equiv \hat{\lambda}(u_*,u_1) \equiv \frac{1}{u_1^i-u_*^i} \int_{\substack{u_*^i \\ u \in R_i(u_*)}}^{u_1^i} \lambda_i(u)\,du^i.$$

Moreover, when $u_* \ne u_-$, $(u_*,\tilde{u}_*)$ is a left-sided contact discontinuity: $\sigma(u_*,\tilde{u}_*) = \lambda_i(u_*)$.

PROOF: When $\theta = 0$, the lemma holds trivially with $u_* = u_1$. Suppose that $\theta > 0$. Given u on $R_i(u_-)$ between u_- and u_1, let $\tilde{u}$ be the point on $S_i(u)$ with $\tilde{u}^i = u_+^i$ and $\theta(u) \equiv \lambda_i(u) - \sigma(u,\tilde{u})$. Thus $\theta(u_1) = \theta > 0$.

Claim 1: So long as $\theta(u)$ is positive, $\theta(u)$ decreases as u moves along $R_i(u_-)$ from u_1 toward u_-.

The claim is proved by the estimate (2.10). We choose μ_i to be u^i and r_i to have unit i-component so that we have from (2.4) and (2.9)

$$\beta_i \equiv \beta_i(\tilde{u},u) = 1 + O(1)|u - \tilde{u}|(\lambda_i - \sigma)$$

$$\beta_j = O(1)|u - \tilde{u}|(\lambda_i - \sigma), \quad j \ne i.$$

Given $\varepsilon > 0$, choose w on $R_i(u_-)$ with $u_-^i - w^i = \varepsilon$. From above we have

(5.14) $$\|\bar{u} - w\| = O(1)|u - \tilde{u}| \varepsilon \theta(u), \quad \bar{u}^i = w^i,$$

for some $\bar{u}$ on $S_i(\tilde{u})$. This, along with (2.10), implies that

(5.15) $\dfrac{\theta(w) - \theta(u)}{\varepsilon} = \lambda_i(w) - \lambda_i(u) + [-\dfrac{1}{\tilde{u}^i - u^i} + O(1)]\theta(u).$

Since $\nabla\lambda_i \cdot r_i > 0$ along $R_i(u_-)$ between u_- and u_1, we have $\lambda_i(w) - \lambda_i(u) < 0$. Therefore (6.14) implies that for $\tilde{u} - u$ weak $\theta(w) < \theta(u)$. This proves the claim.

Claim 2: Let w and u be given as above, and set $\alpha \equiv \tilde{u}^i - u^i > 0$, then

$$(\alpha+\varepsilon)\sigma(\tilde{w},w) = \alpha\sigma(\tilde{u},u) + \varepsilon\lambda_i(u) + O(1)\varepsilon\alpha\theta(u) + O(1)\varepsilon^2.$$

The claim follows from (5.15) and the identities $|\lambda_i(u) - \lambda_i(w)| + |\theta(w) - \theta(u)| = O(1)\varepsilon$ and $\theta \equiv \lambda_i - \sigma$. We omit the tedious calculations. For the same w and u we have the following:

Claim 3: $(w,\tilde{w})$ is admissible provided that $(u,\tilde{u})$ is admissible, $\theta(u) > 0$ and ε is small.

If $(w,\tilde{w})$ is not admissible then there exists $\hat{w} \in S_i(\tilde{w})$ such that $\sigma(\tilde{w},\hat{w}) = \sigma(\tilde{w},w)$. From (5.14) there exists $\hat{u}$ on $S_i(\tilde{u})$ such that $\hat{u}^i = \hat{w}^i$ and

$$\sigma(\tilde{u},\hat{u}) = \sigma(\tilde{w},\hat{w}) + O(1)|u - \tilde{u}|\theta(u)\varepsilon.$$

Since $(u,\tilde{u})$ satisfies condition (E), we have $\sigma(\tilde{u},u) \geq \sigma(\tilde{u},\hat{u})$ and so the above estimates imply

$$\sigma(\tilde{w},w) = \sigma(\tilde{w},\hat{w}) \leq \sigma(\tilde{u},u) + O(1)|u - \tilde{u}|\theta(u)\varepsilon.$$

On the other hand, we have from Claim 2 that

$$\alpha[\sigma(\tilde{w},w) - \sigma(\tilde{u},w)] = \varepsilon\theta(u) + \varepsilon[\sigma(\tilde{u},u) - \sigma(\tilde{w},w)] + O(1)\varepsilon\alpha\theta(u) + O(1)\varepsilon^2$$

$$= \varepsilon[\theta(u) + O(1)\alpha\theta(u) + \varepsilon].$$

The last two estimates yield

$$O(1)|u - \tilde{u}|\theta(u)\varepsilon\alpha \geq \varepsilon[\theta(u) + O(1)\alpha\theta(u) + \varepsilon]$$

which is a contradiction since we assume that $\theta(u)$ is positive, ε is small and waves are weak. This proves Claim 3. To finish the proof of the lemma we see from Claim 1 that there are two cases:

Case 1: $\theta(u_*) = 0$ for some u_* on $R_i(u_-)$ between u_- and u_1 and $\theta(u) > 0$ for all u on $R_i(u_-)$ between u_* and u_1.

Case 2: $\theta(u) > 0$ for all u on $(R_i(u_-)$ between u_- and u_1.

In the second case we set $u_* = u_-$. We will only treat the first case which is harder. As in the statement of the lemma, we set $\alpha_2 \equiv \tilde{u}_*^i - u_*^i$ where $\tilde{u}_*$ is the state on $S_i(u_*)$ with $\tilde{u}_*^i = u_+^i$. (iii) of the lemma follows from Claim 3. To prove (i) and (ii), we choose a monotone sequence of states $u_1, u_2, \ldots, u_m = u_*$ on $R_i(u_-)$ with $u_j^i - u_{j-1}^i = \frac{\alpha_2}{m}$, m any large integer. As before we choose $\tilde{u}_j$ on $S_i(u_j)$ with $\tilde{u}_j^i = u_+^i$. The strength of $(u_j, \tilde{u}_j)$ is $\alpha_1 + \frac{j}{m}\alpha_2$. From (5.14) and (5.15) we have

$$|\tilde{u}_j - u_+| = O(1) \sum_{j=1}^{m} (\alpha_1 + \frac{j}{m}\alpha_2) \frac{\alpha_2}{m} \theta(u_j)$$

$$(\alpha_1 + \alpha_2)\sigma = \alpha_1\sigma_1 + \alpha_2 \frac{1}{m} \sum_{j=1}^{m} \lambda_i(u_j) + O(1) \frac{\alpha_2}{m} \sum_{j=1}^{m} (\alpha_1 + \frac{j}{m}\alpha_2)\theta(u_j) + O(1) \frac{(\alpha_2)^2}{m^2}.$$

If we can show that for any large m

(5.16) $$\frac{\alpha_2}{m} \sum_{j=1}^{m} (\alpha_1 + \frac{j}{m}\alpha_2)\theta(u_j) = O(1) \theta\beta \min(\alpha_1, \alpha_2)$$

then (i) and (ii) of the lemma will be the consequence of the above two estimates as $m \to \infty$. We have from (5.15) that

(5.17) $$0 \leq \theta(u_{j+1}) \leq \theta(u_j) + \frac{\alpha_2}{m}[-(\alpha_1 + \frac{j\alpha_2}{m})^{-1} + O(1)]\theta(u_j)$$

$$= \frac{m\alpha_1 + (j-1)\alpha_2}{m\alpha_1 + j\alpha_2} \theta(u_j) + O(1) \frac{\alpha_2}{m} \theta(u_j).$$

We know that $\theta(u_1) = \theta$. The above estimate inductively implies that

$$\theta(u_{j+1}) = (\prod_{k=1}^{j} \frac{m\alpha_1 + (k-1)\alpha_2}{m\alpha_1 + k\alpha_2})\theta + O(1) \frac{\alpha_2}{m} \sum_{k=1}^{j} \theta(u_k)$$

$$= \frac{m\alpha_1}{m\alpha_1 + j\alpha_2} \theta + O(1)\alpha_2\theta$$

$$\frac{\alpha_2}{m} \sum_{j=1}^{m} (\alpha_1 + \frac{j}{m}\alpha_2)\theta(u_j) = \alpha_1\alpha_2\theta + O(1)\beta(\alpha_2)^2\theta.$$

This proves (5.16) when $\alpha_2 \leq \alpha_1$. Next, suppose that $\alpha_2 > \alpha_1$. Choose integer M with

(5.18) $(M+1)\alpha_1 \geq \alpha_2 \geq M\alpha_1$

and let $m = Mn$, n any large integer. We have from (5.17) and the second inequality in (5.18) that

$$\theta(u_{j+1}) \leq \left[\frac{Mn\alpha_1+(j-1)\alpha_2}{Mn\alpha_1+j\alpha_2} + O(1)\frac{\alpha_2}{nM} \right] \theta(u_j)$$

$$\leq \left[\frac{n+j-1}{n+j} + O(1)\frac{\alpha_2}{nM} \right] \theta(u_j), \quad j = 1,2,\ldots,Mn.$$

From the basic assumption that $O(1)\alpha_2$ is small we have

$$O(1)\frac{\alpha_2}{nM} \leq \frac{1}{6nM}$$

and so, direct calculations yield

$$\theta(u_{j+1}) \leq \left[\frac{2n+j-1}{2n+j} \right] \theta(u_j)$$

$$\leq \frac{2n}{2n+j}\theta \quad \text{for } j = 0,1,2,\ldots,(n-1)M,$$

$$\theta(u_j) \leq \frac{2n}{2n+(n-1)M}\theta \quad \text{for } j = (n-1)M,\ldots,nM.$$

Finally, the above estimates are used to verify (5.16):

$$\frac{\alpha_2}{m}\sum_{j=1}^{m}(\alpha_1+\frac{j}{m}\alpha_2)\theta(u_j) \leq \frac{2\alpha_1\alpha_2\theta}{n^2M}\left[\sum_{j=1}^{(n-1)M}\frac{2nj}{2n+j} + \sum_{j=(n-1)M}^{nM}\frac{2n^2M}{2n+(n-1)M} \right]$$

$$\leq \frac{4\alpha_1\alpha_2\theta}{nM}\sum_{j=1}^{(n-1)M}\frac{j}{2n+j} + 4\alpha_1\alpha_2\theta\sum_{j=(n+1)M}^{nM}\frac{1}{2n+1}(n-1)M$$

$$\leq 4\alpha_1\alpha_2\theta .$$

This completes the proof of the lemma.   Q.E.D.

§6.   Stability of wave pattern

The first result, Lemma 6.3, states that elementary i-waves contained in the solution of a Riemann problem constructed in Section 4 depends smoothly on the end states not only in their strengths and speeds, but also in their wave types. Suppose that $u_+ \in T_i(u_-)$ so that u_- is related to u_+ by i-discontinuities (u_{j-1},u_j), and i-rarefaction waves (u_j,u_{j+1}), j odd, $1 \leq j \leq m$, $u_0 = u_-$, $u_m = u_+$. We set

(6.1)
$$\theta_\ell(u_-,u_+) \equiv \lambda_i(u_-) - \sigma(u_-,u_1) \geq 0,$$

$$\theta_r(u_-,u_+) \equiv \sigma(u_{m-1},u_+) - \lambda_i(u_+) \geq 0.$$

When u_ℓ and u_r are related by an i-discontinuity, we set

$$D_i(u_\ell,u_r) \equiv \{u: (u-u_\ell)\sigma(u_\ell,u_r) - (f(u)-f(u_\ell))$$

$$= c(u)r_i(u) \text{ for some scalar } c(u)\}.$$

A state u is called <u>between</u> an admissible i-discontinuity (u_ℓ,u_r) if $u \in D_i(u_\ell,u_r)$ between u_ℓ and u_r, and u is <u>between</u> an i-rarefaction wave (u_ℓ,u_r) if $u \in R_i(u_\ell,u_r)$ between u_ℓ and u_r.

The following lemma is an easy consequence of the strict hyperbolicity of the system (1.1) (c.f. Lemmas 2.1 and 2.5). Its proof is omitted.

<u>LEMMA</u> 6.1. For u_ℓ and u_r close, $D_i(u_\ell,u_r)$ is a smooth curve through u_ℓ and u_r in a small neighborhood of u_r; (u_ℓ,u_r) is admissible if $c(u)$ is nonnegative for u between u_ℓ and u_r. (Here for definiteness, $r_i(u)$ is chosen to be in the same direction as $u_r - u_\ell$.) Moreover, if a state $\tilde{u}$ satisfies

$$(\tilde{u}-u_\ell)\sigma - (f(\tilde{u})-f(u_\ell)) = \tilde{c}r_i(\tilde{u}) + K$$

for some scalars $\tilde{c}$ and for some vector K, then there exists a vector u on $D_i(u_\ell,u_r)$ such that

$$\|u - \tilde{u}\| = O(1)\|K\|.$$

<u>Definition</u> 6.1. A set of vectors $\{v_0,v_1,\ldots,v_\ell\}$ is a <u>partition</u> of the i-waves in (u_-,u_+) if

(i) $v_0 = u_-,$ $v_\ell = u_+,$ $v_{k-1}^i \leq v_k^i,$ $k = 1,2,\ldots,\ell,$

(ii) $\{v_0,v_1,\ldots,v_\ell\} \supset \{u_0,u_1,\ldots,u_m\},$

(iii) $v_k \in R_i(u_j),$ j odd, if $u_j^i < v_k^i \leq u_{j+1}^i,$

(iv) $v_k \in D_i(u_{j-1},u_j),$ j odd, if $u_{j-1}^i < v_k^i \leq u_j^i.$

We set

(v) $Y_k \equiv v_k - v_{k-1},$

(vi) $\lambda_{i,k} \equiv \lambda_i(v_{k-1})$ and $[\lambda_i]_k \equiv [\lambda_i](v_{k-1},v_k) \equiv \lambda_i(v_k) - \lambda_i(v_{k-1}) > 0$ if (iii) holds.

(vii) $\lambda_{i,k} \equiv \sigma(u_{j-1},u_j)$ and $[\lambda_i]_k \equiv [\lambda_i](v_{k-1},v_k) \equiv 0$ if (iv) holds.

A partition $\{w_r\}$ is _finer_ than another partition $\{v_k\}$ if $\{w_k\} \supset \{v_k\}$. In the lemmas or theorems that follow, when a result is said to hold for a partition then it also holds for any finer partition.

LEMMA 6.3. Suppose that $u_+ \in T_i(u_-)$, $\bar{u}_+ \in T_i(\bar{u}_-)$, $u_+^i - u_-^i = \bar{u}_+^i - \bar{u}_-^i \equiv \alpha > 0$, and $\|u_- - \bar{u}_-\| \equiv \beta$. Then there exists a partition $\{v_0, v_1, \ldots, v_\ell\}$ for the i-waves relating (u_-, u_+) such that there is a partition $\{\bar{v}_0, \bar{v}_1, \ldots, \bar{v}_\ell\}$ for the i-waves relating $(\bar{u}_-, \bar{u}_+)$ such that $\bar{v}_k^i - \bar{v}_0^i = v_k^i - v_0^i$, $k = 1, 2, \ldots, \ell$, and

(i) $\sum_{k=0}^{\ell} \|y_k - \bar{y}_k\| = O(1)\alpha\beta$,

(ii) $|\lambda_{i,k} - \bar{\lambda}_{i,k}| = O(1)\beta$, $k = 0, 1, 2, \ldots, \ell$,

(iii) $|\theta_\ell(u_-, u_+) - \theta_\ell(\bar{u}_-, \bar{u}_+)| + |\theta_r(u_-, u_+) - \theta_r(\bar{u}_-, \bar{u}_+)| = O(1)\alpha\beta$.

Moreover, $\{1, 2, \ldots, \ell\}$ can be written as a disjoint union of subsets I, II and III such that

(iv) for $k \in$ I, both v_k and $\bar{v}_k$ are of type (iii) of Definition 6.2 and

$$\sum_{k \in I} |[\lambda_i]_k - [\bar{\lambda}_i]_k| = O(1)\alpha\beta,$$

(v) for $k \in$ II, both v_k and $\bar{v}_k$ are of type (iv) of Definition 6.2,

(vi) for $k \in$ III, v_k and $\bar{v}_k$ are of different types and

$$\sum_{k \in III} |[\lambda_i]_k + [\bar{\lambda}_i]_k| = O(1)\alpha\beta.$$

PROOF: We first partition (u_-, u_+) and $(\bar{u}_-, \bar{u}_+)$ separately. By refining these partitions, we obtain easily $\{v_k\}$ and $\{\bar{v}_k\}$ with $v_k^i - v_-^i = \bar{v}_k^i - \bar{v}_-^i$ for all k. Moreover, when $\{v_k\}$ is refined, $\{\bar{v}_k\}$ can be easily refined accordingly. The division of $\{1, 2, \ldots, \ell\}$ into I, II and III is obvious. We first prove the lemma when β is small. Let u_- (or u_+) move along the line between u_- and $\bar{u}_-$ (or between u_+ and $\bar{u}_+$). For β sufficiently small, an i-discontinuity moves to a corresponding i-discontinuity $(\bar{u}_{j-1}, \bar{u}_j)$ in $(\bar{u}_-, \bar{u}_+)$. Of course, (u_{j-1}, u_j) (or $(\bar{u}_{j-1}, \bar{u}_j)$) may join other continuity in (u_-, u_+) (or $(\bar{u}_-, \bar{u}_+)$ and be part of another discontinuity, say, (u_{j-1}, u_{j+1}). If this is the case, we choose β small so that $\sigma(u_{j-1}, u_{j+1}) = \sigma(u_{j-1}, u_j) = \lambda_i(u_j)$, and we still regard (u_{j-1}, u_{j+1}) as two discontinuities. In general, for small β, (u_{j-1}, u_j) and $(\bar{u}_{j-1}, \bar{u}_j)$ are isolated in the sense that there exist w_k and $\bar{w}_k$, $k = 1, 2, 3, 4$, which form part of partitions for (u_-, u_+) and

$(\overline{u}_-, \overline{u}_+)$ respectively, $\{w_k^i - w_-^i, \ k = 1,2,3,4\} = \{\overline{w}_k^i - \overline{w}_-^i, \ k = 1,2,3,4\} =$
$\{u_{j-1}^i - u_-^i, \ u_j^i - u_-^i, \ \overline{u}_{j-1}^i - \overline{u}_-^i, \ \overline{u}_j^i - \overline{u}_-^i\}$, and (u_{j-1}, u_j) $((\overline{u}_{j-1}, \overline{u}_j))$ is
is the only discontinuity in (w_1, w_4) $((\overline{w}_1, \overline{w}_4))$. For simplicity, we will
treat only the case when (u_{j-1}, u_j) and $(\overline{u}_{j-1}, \overline{u}_j)$ are isolated in the
above sense; similar principles apply to other cases. We now prove that
the lemma holds for the partitions for (u_-, w_4) and $(\overline{u}_-, \overline{w}_4)$ assuming
that it already holds for the partitions for (u_-, w_1) and $(\overline{u}_-, \overline{w}_1)$. Thus
we assume, as an induction hypothesis, that

(6.2) $$\|w_1 - \overline{w}_1\| = O(1)\beta .$$

There are several different ways how $\{w_1, w_2, w_3, w_4\}$ is related to
$\{u_{j-1}, u_j\}$. We will only treat the case when $u_{j-1} = w_2$ and $u_j = w_4$,
Figure 6.1; other cases are treated in a similar way.

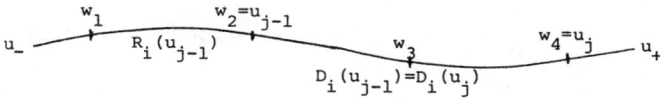

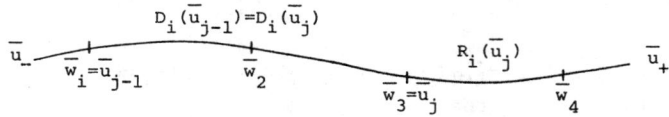

Figure 6.1

To prove (vi) we have to show that

(6.3) $$\lambda_i(w_2) - \lambda_i(w_1) + \lambda_i(\overline{w}_4) - \lambda_i(\overline{w}_3) = O(1)\beta\|w_1 - w_4\| .$$

When $w_4 = u_+$ and $\overline{w}_3 = \overline{w}_4 = \overline{u}_+$, (iii) becomes

(6.4) $$\sigma(u_{j-1}, u_j) - \lambda_i(u_j) = \sigma(\overline{u}_{j-1}, \overline{u}_j) - \lambda_i(\overline{u}_j) + O(1)\beta|w_1^i - w_4^i| .$$

For simplicity we assume that $\sigma(u_{j-1}, u_j) = \lambda_i(u_{j-1})$ and $\sigma(\bar{u}_{j-1}, \bar{u}_j) = \lambda_i(\bar{u}_{j-1})$. This is the case if $w_1 \neq u_-$; the case where $w_1 = u_-$ is treated using the same arguments. We have the following two cases:

Case 1: $\bar{w}_3 \neq \bar{w}_+$, $w_3 \neq w_+$ and $\sigma(\tilde{w}_1, \tilde{w}_3) = \lambda_i(\tilde{w}_3)$.

Case 2: $w_3 = w_4 = w_+$.

In the first case we pick $\bar{w}_*$ on $S_i(\bar{w}_1)$ with $\bar{w}_*^i = \bar{w}_4^i$. Since $\sigma(\bar{w}_1, \bar{w}_3) = \lambda_i(\bar{w}_3)$ we have

$$(6.5) \quad (\bar{w}_4^i - \bar{w}_1^i)\sigma(\bar{w}_1, \bar{w}_*) = (\bar{w}_3^i - \bar{w}_1^i)\sigma(\bar{w}_3, \bar{w}_1) + (\bar{w}_4^i - \bar{w}_3^i)\hat{\lambda}_i(\bar{w}_3, \bar{w}_4)$$
$$+ O(1)|\bar{w}_3^i - \bar{w}_4^i| \cdot |\bar{w}_1^i - \bar{w}_4^i| \cdot (\lambda_i(\bar{w}_4) - \lambda_i(\bar{w}_3)).$$

The above estimate is proved using the same techniques in establishing (ii) of Lemma 5.2; details are omitted. Similarly, since $\sigma(w_2, w_4) = \lambda_i(w_2)$, we have

$$(6.6) \quad (w_4^i - w_1^i)\sigma(w_1, w_*) = (w_4^i - w_2^i)\sigma(w_2, w_4) + (w_2^i - w_1^i)\hat{\lambda}_i(w_1, w_2)$$
$$+ O(1)|w_4^i - w_1^i| \cdot |w_2^i - w_1^i| \cdot (\lambda_i(w_2) - \lambda_i(w_1))$$

where w_* is a point on $S_i(w_1)$ with $w_*^i = w_4^i$. We have from continuity arguments and (6.2) that

$$(6.7) \quad \sigma(w_1, w_*) - \lambda_i(w_1) = \sigma(\bar{w}_1, \bar{w}_*) - \lambda_i(\bar{w}_1) + O(1)\beta|w_4^i - w_1^i|.$$

It follows from identities $\sigma(w_2, w_4) = \lambda_i(w_2)$, $\sigma(\tilde{w}_1, \tilde{w}_3) = \lambda_i(\tilde{w}_1) = \lambda_i(\tilde{w}_3)$ and estimates $(6.5) \sim (6.7)$ that

$$(w_4^i - w_2^i)(\lambda_i(w_2) - \lambda_i(w_1)) + (w_2^i - w_1^i)(\hat{\lambda}_i(w_2, w_1) - \lambda_i(w_1)) + (w_4^i - w_3^i)(\lambda_i(\tilde{w}_3) - \hat{\lambda}_i(\bar{w}_3, \bar{w}_4))$$

$$= O(1)|w_4^i - w_1^i|\{|w_4^i - w_1^i| \cdot \beta + |w_2^i - w_1^i|(\lambda_i(w_2) - \lambda_i(w_1)) + |w_4^i - w_3^i|(\lambda_i(\bar{w}_4) - \lambda_i(\bar{w}_3))\},$$

$$(\bar{w}_3^i - \bar{w}_1^i)(\lambda_i(\bar{w}_4) - \lambda_i(\bar{w}_3)) + (\bar{w}_4^i - \bar{w}_3^i)(\hat{\lambda}_i(\bar{w}_3, \bar{w}_4) - \lambda_i(\bar{w}_3)) + (\bar{w}_2^i - \bar{w}_1^i)(\lambda_i(w_1) - \hat{\lambda}_i(w_2, w_1))$$

$$= O(1)|\bar{w}_4^i - \bar{w}_1^i|\{|\bar{w}_4^i - \bar{w}_1^i|\beta + |w_2^i - w_1^i|(\lambda_i(w_2) - \lambda_i(w_1)) + |w_4^i - w_3^i|(\lambda_i(\bar{w}_4) - \lambda_i(\bar{w}_3))\}.$$

Summing up the above two estimates we have

$$\lambda_i(w_2) - \lambda_i(w_1) = (1 + O(1)|w_4^i - w_1^i|)(\lambda_i(\bar{w}_4) - \lambda_i(\bar{w}_3)) + O(1)|w_4^i - w_1^i|\beta.$$

Notice that $\lambda_i(w_1) \le \hat{\lambda}_i(w_1,w_2) \le \lambda_i(w_2)$ and $\lambda_i(\overline{w}_3) \le \hat{\lambda}_i(\overline{w}_3,\overline{w}_4) \le \lambda_i(\overline{w}_4)$. Moreover,, for small β, we may assume that $|w_4^i - w_1^i| \le 2|w_3^i - w_2^i|$. These inequalities, along with the above three estimates, yield (6.3).

We now consider Case 2. Estimate (6.3) is obtained as above. As a consequence of (6.3) (c.f.(6.5) and (6.6)), we have

$$(6.8) \qquad \sigma(w_1,w_\star) = \sigma(w_2,w_4) + O(1)\beta|w_4^i - w_1^i|$$

$$\sigma(\overline{w}_1,\overline{w}_\star) = \sigma(\overline{w}_1,\overline{w}_3) + O(1)\beta|w_+^i - w_1^i|$$

$$\|w_\star - w_4\| + \|\overline{w}_\star - \overline{w}_4\| = O(1)\beta|w_4^i - w_1^i|.$$

From continuity arguments, we have

$$\sigma(w_1,w_\star) - \lambda_i(w_\star) = \sigma(\overline{w}_1,\overline{w}_\star) - \lambda_i(\overline{w}_\star) + O(1)\beta|w_4^i - w_1^i|.$$

The above estimates, along with the identities $\sigma(w_2,w_4) = \lambda_i(w_2)$ and $\sigma(\overline{w}_1,\overline{w}_3) = \lambda_i(\overline{w}_1)$, imply (6.4).

We now prove (i) and (ii) of the lemma. For brevity, we will only show that for w on $D_i(w_2,w_4)$ and $\overline{w}$ on $D_i(\overline{w}_1,\overline{w}_3)$, $w^i - w_-^i = \overline{w}^i - \overline{w}_-^i$ between $w_3^i - w_-^i$ and $w_3^i - w_-^i$,

$$(6.9) \qquad w - w_1 = \overline{w} - \overline{w}_1 + O(1)\beta|w^i - w_1^i|.$$

Pick $\hat{w}_3$ on $S_i(w_1)$ with $\hat{w}_3^i = w_3^i$. From continuity arguments we see that for any $\hat{w}$ on $D_i(w_1,\hat{w}_3)$ and $\overline{w}$ on $D_i(\overline{w}_1,\overline{w}_3)$ with $\hat{w}^i - w_1^i = \overline{w}^i - \overline{w}_1^i$

$$(6.10) \qquad \hat{w} - w_1 = \overline{w} - \overline{w}_1 + O(1)\beta|\overline{w}^i - \overline{w}_1^i|$$

$$(6.11) \qquad \sigma(w_1,\hat{w}_3) - \lambda_i(w_1) = \sigma(\overline{w}_1,\overline{w}_3) - \lambda_i(\overline{w}_1) + O(1)\beta|w_3^i - w_1^i|.$$

It follows from (6.3), (6.7), (6.8) and (6.11) that

$$(6.12) \qquad \sigma(w_1,\hat{w}_3) = \sigma(w_2,w_4) + O(1)\beta|w_1^i - w_4^i|$$

$$= \lambda_i(w_2) + O(1)\beta|w_1^i - w_4^i|.$$

We now compare $\hat{w}$ with w. It follows from (6.3) and (6.12) that

(6.13)

$$= (\hat{w}-w_1)\sigma(w_1,\hat{w}_3) - (f(\hat{w}) - f(w_1)) + (w_1-w_2)\lambda_i(w_2) + f(w_2) - f(w_1)$$

$$+ O(1)\beta|w_1^i - w_4^i||\hat{w}^i - w_1^i|.$$

Since $\hat{w} \in D_i(w_i,\hat{w}_3)$, we have

(6.14) $$(\hat{w}-w_1)\sigma(w_1,\hat{w}_3) - (f(\hat{w}) - f(w_1)) = cr_i(\hat{w})$$

for some scalar c. From (6.3) one can show that for $\hat{w}_2$ on $S_i(w_1)$ with $\hat{w}_2^i = w_2^i$, we have

$$\sigma(w_1,\hat{w}_2) = \lambda_i(w_2) + O(1)\beta|w_4^i - w_1^i|$$

$$\|\hat{w}_2 - \hat{w}_1)\| = O(1)\beta|w_4^i - w_1^i||w_1^i - w_2^i|,$$

(c.f. Lemmas 5.2 and (6.5)). These two estimates imply that

$$(w_1-w_2)\lambda_i(w_2) + f(w_2) - f(w_2) = O(1)\beta|w_4^i - w_1^i||w_4^i - w_1^i|.$$

The above estimates and (6.13), (6.14) yield that

$$(\hat{w}-w_2)\sigma(w_2,w_4) - (f(\hat{w})-f(w_2)) = O(1)\beta|w_4^i - w_1^i||w_4^i - w_1^i|$$

whence, by Lemma 6.1, we have (6.9).

For simplicity we do not treat all possible cases in the above proof. However, similar arguments can be applied to other cases to show that the lemma holds for sufficiently small β. For β not small enough, we choose sequences $u_- = u_{1-}, u_{2-}, \ldots, u_{m-} = \tilde{u}_-$ between u_- and $\tilde{u}_-$, $|u_{j-} - u_{(j+1)-}|$ small, $j = 1,2,\ldots,m-1$, and apply the above result repeatedly to (u_{j-},u_{j+}) and $(u_{(j+1)-},u_{(j+1)+})$. Here u_{j+} is a state on $T_i(u_{j-})$ with $u_{j+}^i - u_{j-}^i = u_+^i - u_-^i$. When (u_-,u_+) and (u_{j-},u_{j+}) have been partitioned so that the lemma holds for them, we apply the above result to (u_{j-},u_{j+}) and $(u_{(j+1)-},u_{(j+1)+})$ by refining the partition for (u_{j-},u_{j+}). Thus in each step we refine the partition for (u_-,u_+). This completes the proof of the lemma.   Q.E.D.

The above lemma shows that the wave pattern in the solution of a Riemann problem changes smoothly with respect to its boundary states. Our next lemma shows that initial states change smoothly with respect to wave

patterns. The elementary waves in a solution of the Riemann problem is
noninteracting in the sense that none of these waves tend toward each other.
A wave pattern consisting of elementary waves is said to be weakly interacting if any two of these waves make a small angle between them. The
next lemma shows that a weakly interacting wave pattern is close to the
wave pattern in a solution of the Riemann problem. This is a nonlinear
phenomena and is clearly not true for linear system for which any wave pattern is noninteracting. For convenience, we classify discontinuity waves
into the following two kinds. An admissible i-discontinuity (u_-,u_+) is
said to be simple if $\sigma(u_-,u) \neq \sigma(u_-,u_+)$ for any u on $S_i(u_-)$ between
u_- and u_+. When there exist $u_1,u_2,\ldots,u_m$ on $S_i(u_-)$ between u_- and
u_+ such that $\sigma(u_-,u_i) = \sigma(u_-,u_+)$, $i = 1,2,\ldots,m$, then it is clear from
the jump condition that $u_{i+1} \in S_i(u_i)$ and (u_i,u_{i+1}) is admissible. In
this case, we say that (u_-,u_+) is a composite of (u_i,u_{i+1}), $i =$
$1,2,\ldots,m-1$. A small perturbation of the end states u_- and u_+ may
split a composite discontinuity (u_-,u_+) into weaker discontinuities. Thus
the stability of the wave pattern has to be understood in the above sense.
For genuinely nonlinear system, all discontinuity waves are simple and the
wave pattern has a stronger stability property. We expect, however, an
elementary discontinuity is generically simple.

LEMMA 6.4. Suppose that u_0 and u_m are related by weak i-elementary
waves (u_{j-1},u_j), $j = 1,2,\ldots,m$, $m \geq 2$, with finite total strength
$\sum_{j=1}^m \|u_{j-1}-u_j\| = M < \infty$. Assume that the following holds for some positive
constant ε:

  (a) if (u_0,u_1) (or (u_{m-1},u_m) is an i-discontinuity, then
       $0 \leq \sigma(u_0,u_1) - \lambda_i(u_1) \leq \varepsilon$ (or $0 \leq \lambda_i(u_{m-1}) - \sigma(u_{m-1},u_m) \leq \varepsilon$),

  (b) $\lambda_i(u_{j_1}) - \lambda_i(u_{j_2}) \leq \varepsilon$ for any j_1,j_2, $1 \leq j_1 \leq j_2 < m$.

Then there exists a state u_* on $T_i(u_0)$ such that $u_*^i = u_m^i$ and

  (i) $\|u_m-u_*\| = O(1)\varepsilon$.

Here $O(1)$ is independent of the integer $m = m(\varepsilon)$. In particular, if
$u_0 = u_0(\varepsilon) \to u_\ell$ and $u_m = u_m(\varepsilon) \to u_r$ as $\varepsilon \to 0$, then $u_r \in T_i(u_\ell)$.
Moreover $\{(u_{j-1},u_j)\}$ is close to the i-waves relating u_ℓ and u_r
in the following sense: Suppose that u_ℓ is related to u_r
by i-discontinuities (w_{k-1},w_k), k odd, and i-rarefaction waves
(w_{k-1},w_k), k even, $k = 1,2,\ldots,\ell$, as constructed in Section 4. Then for
ε sufficiently small, $0 < \varepsilon \leq \varepsilon_0$, ε_0 depending on the i-waves relating
u_ℓ and u_r, there exists a strictly monotone sequence $\{j_1,j_2,\ldots,j_\ell\}$
$\subset \{1,2,\ldots,m\}$ such that

Since $R_i(u_0^{\hat{}})$ is transversal to Ω_0^i, the above estimate imply that for sufficiently small ε, all states between (u_{j-1}, u_j), $j = 1,2,\ldots,m$, take values in a small neighborhood of u_0 with diameter $0(\varepsilon)$ (c.f. (3.2)). Thus the lemma holds trivially.

In Case 2, an i-discontinuity (u_{j-1}, u_j) either lies outside $\Omega_{\varepsilon}1/3$ or lies inside $\Omega_{2\varepsilon}1/3$ or both. In the formal case, the strength of (u_{j-1}, u_j) is $0(1)(\varepsilon^{1/3})$ as a consequence of hypothesis (b) and (i) of Lemma 2.4. Consequently, we have from (ii) of Lemma 2.4 that all states between (u_{j-1}, u_j), $j = 1,2,\ldots,m$, take values in $R_i(u_0) + E'$ where the error term E' satisfies

$$E' = 0(1)[(\sqrt{\varepsilon}) + 2\sqrt{\varepsilon} \; \phi_i(2\sqrt{\varepsilon})] \sum_{j=1}^{m} \|u_{j-1} - u_j\|$$

$$= 0(1)M\sqrt{\varepsilon} \; .$$

This estimate, along with hypothesis (b) and (i) of Lemma 2.4, implies that if two states $\bar{u}$ and $\bar{\bar{u}}$, each between an i-wave in $\{(u_{j-1}, u_j)$, $j = 1,2,\ldots,m\}$, lies outside $\Omega_{2\sqrt{\varepsilon}}$, then $\nabla\lambda_i \cdot r_i(\bar{u})$ and $\nabla\lambda_i \cdot r_i(\bar{\bar{u}})$ have the same sign. Moreover, $R_i(u_0)$ outside $\Omega_{2\sqrt{\varepsilon}}^i$ can be partitioned so that analogous estimate as (ii) in the lemma holds. Consequently, there exists rarefaction waves taking values along $R_i(u_0) \cap (\Omega_{2\sqrt{\varepsilon}}^i)^c$ with the property that u_0 is related to u_m monotonically (mod $0\sqrt{\varepsilon}$)) by these rarefaction waves and i-waves in $\Omega_{2\sqrt{\varepsilon}}$. We note that from Case 1 the i-waves in $\Omega_{2\sqrt{\varepsilon}}$ stay in small neighborhoods of finite points in $\Omega_{2\sqrt{\varepsilon}}$. Finally, we use Lemma 6.3 on the stability of wave pattern to prove the lemma. Details are omitted.

We now treat Case 3. Let (u_{j-1}, u_j) be an i-discontinuity with the property stated in Case 3. For definiteness, we assume that $u_j - u_{j-1}$ is in the direction of r_i which, for convenience, we call the direction to the right. We only treat the case when $j-1 \neq 0$ and $j \neq m$; other cases can be treated similarly. Thus our hypothesis (a) and (b) imply that

$$(6.15)_1 \qquad\qquad 0 \leq \sigma(u_{j-1}, u_j) - \lambda_i(u_j) \leq \varepsilon,$$

$$(6.15)_2 \qquad\qquad 0 \leq \lambda_i(u_{j-1}) - \sigma(u_{j-1}, u_j) \leq \varepsilon.$$

Suppose that between u_{j-1} and u_j $S_i(u_{j-1})$ lies in $\Omega_{2\varepsilon}1/3$. In this case $S_i(u_{j-1})$ differs from $D_i(u_{j-1}, u_j)$ and $R_i(u_{j-1})$ by $0(1)\sqrt{\varepsilon} \; \phi(r\sqrt{\varepsilon})$ (c.f. Lemma 2.4). Thus $\bar{u}$ and $\bar{\bar{u}}$ in Case 3 do not exist. Therefore, if $u_{j-1} \in \Omega_{\varepsilon}1/3$ there exists u'' at which $S_i(u_{j-1})$ first leaves $\Omega_{2\varepsilon}1/3$ as u moves along $S_i(u_{j-1})$ from u_{j-1} toward u_j. Denote by u' the last point on $S_i(u_{j-1}) \cap \Omega_{\varepsilon^{1/3}}$ between u_{j-1} and u''. We have from the proof of Lemma 3.4 that if $\nabla \lambda_i \cdot r_i(u) < 0$ for u on $S_i(u_{j-1})$ between u_{j-1} and u_j, then either for some $C_0 > 0$

$$(6.16)_1 \qquad\qquad \sigma(u_{j-1},u') - \lambda_i(u') \leq -C_0\varepsilon^{2/3}$$

or $\qquad \dfrac{d\sigma(u_{j-1},u)}{d\mu_i} < 0$ at $u = u''$ and for some $C_1 > 0$

$$(6.16)_2 \qquad\qquad \sigma(u_{j-1},u'') - \sigma(u_{j-1},u') \leq -C_1\varepsilon^{2/3}.$$

Since (u_{j-1},u_j) satisfies condition (E), we have from $(6.15)_2$ that

$$(6.17)_1 \qquad\qquad \sigma(u_{j-1},u') \geq \sigma(u_{j-1},u_j) = \lambda_i(u_{j-1}) + C_2\varepsilon$$

$$(6.17)_2 \qquad\qquad \sigma(u_{j-1},u'') \geq \sigma(u_{j-1},u_j) = \lambda_i(u_{j-1}) + C_2\varepsilon$$

for some constant C_2, $0 \leq C_2 \leq 1$. Thus when $(6.16)_1$ holds we have from $(6.17)_1$ that

$$\lambda_i(u') - \lambda_i(u_{j-1}) \geq C_0\varepsilon^{2/3} + C_2\varepsilon$$

and so for sufficiently small ε,

$$(6.18)_1 \qquad\qquad \lambda_i(u') - \lambda_i(u_{j-1}) \geq C_3\varepsilon^{2/3}$$

for some positive constant C_3. Similarly, when $(6.16)_2$ hold, we have from $(6.17)_2$ that

$$\sigma(u_{j-1},u') - \lambda_i(u_{j-1}) \geq C_4\varepsilon^{2/3}$$

for some $C_4 > 0$. The above estimate clearly yields that either $\sigma(u_{j-1},u)$ is increasing at $u = u'$ or there exists $\bar{u}$ on $S_i(u_{j-1})$ between u_{j-1} and u' such that

$$\sigma(u_{j-1},\bar{u}) > \sigma(u_{j-1},u') \quad \text{and} \quad \left.\frac{d\sigma(u_{j-1},u)}{d\mu_i}\right|_{u=\bar{u}} = 0.$$

In either case, we have from Lemma 2.3 that for some $\bar{u}$ between u_{j-1} and u'

(6.18)$_2$ $\lambda_i(\bar{u}) - \lambda_i(u_{j-1}) \geq c_4 \epsilon^{2/3}.$

We have thus proved that one of the following holds

(α) u_{j-1} does not belong to $\Omega_{\epsilon^{1/3}}$ and $\nabla\lambda_i \cdot r_i(u_{j-1}) \leq -\epsilon^{1/3}$,

(β) u_{j-1} belongs to $\Omega_{\epsilon^{1/3}}$ and there exist u', u'' on $S_i(u_{j-1})$
        between u_{j-1} and u_j, with $\nabla\lambda_i \cdot r_i(u') = -\epsilon^{1/3}$, $\nabla\lambda_i \cdot r_i(u'') =$
        $-2\epsilon^{1/3}$, and moreover $\nabla\lambda_i \cdot r_i(u) < -\epsilon^{1/3}$ for all u on $S_i(u_{j-1})$
        between u' and u'', and $u \in \Omega_{2\epsilon^{1/3}}$ for all u on $S_i(u_{j-1})$
        between u_{j-1} and u'',

(γ)   There exists $\bar{u}$ on $S_i(u_{j-1})$ between u_{j-1} and u_j such that
        $|\nabla\lambda_i \cdot r_i(u)| \leq \epsilon^{1/3}$ for all u on $S_i(u_{j-1})$ between u_{j-1} and $\bar{u}$
        and $\lambda_i(\bar{u}) - \lambda_i(u_{j-1}) \geq c_4 \epsilon^{2/3}$ for some positive constant
        c_4.

Analogous statements also hold for the state u_j. We now finish the proof
of the lemma using (α), (β), (γ) and the discussion in Case 2. Suppose
that (u_{j_1-1}, u_{j_1}) is another i-discontinuity with the properties stated in
Case 3, and so $(u_{j'-1}, u_{j'})$, $j_1 < j' < j$, has such properties. Thus the
set $\{(u_{j'-1}, u_{j'}), \ j_1 < j' < j\}$ satisfies conditions of Case 1 and Case 2
and so u_j is related to u_{j_1-1} by i-rarefaction wave (mod $\epsilon^{1/3}$) in the
sense discussed previously. Moreover, it follows from (α), (β) and (γ) that
u_{j-1} lies to the right of u_{j_1} (mod $\epsilon^{1/3}$). This implies that u_{j_1} lies to the
right of u_{j_1-1}. This proves the lemma when all i-discontinuities in (u_ℓ, u_r)
are simple. General situation can be treated similarly using Lemma 6.3.   Q.E.D.

§7.   Interactions of elementary waves II

LEMMA 7.1. Suppose that $u_m \in T_i(u_\ell)$, $u_r \in T_i(u_m)$ and $u_r' > u_m' > u_\ell'$.
Then there exists $u_* \in T_i(u_\ell)$ such that $u_*^i = u_r^i$ and

    (i) $\|u_* - u_r\| = O(1)\,\theta\alpha\beta$

where $\theta = \theta_r(u_\ell, u_m) + \theta_\ell(u_m, u_r)$ (c.f. (6.1)) is the angle between (u_ℓ, u_m)
and (u_m, u_r) and $\alpha = |u_m^i - u_\ell^i|$ and $\beta = |u_r^i - u_m^i|$ are the strength of
(u_ℓ, u_m) and (u_m, u_r) respectively. Moreover, there exist partitions
$\{v_0, v_1, \ldots, v_{j-1}, v_j, \ldots, v_{j+p}\}$ $\{w_{-q}, \ldots, w_0, w_1, \ldots, w_k\}$ and $\{\bar{v}_0, \bar{v}_1, \bar{v}_{j-1}, \bar{w}_1,$
$\bar{w}_2, \ldots, \bar{w}_k\}$ for (u_0, u_m), (u_m, u_r) and (u_ℓ, u_*) respectively, and a
constant $C_{\lambda_i} \equiv C_{\lambda_i}(u, u_m, u_r) \geq 0$, the amount of cancellation of

i-expansion waves, with the following properties:

(ii) $\displaystyle c_{\lambda_i} = \sum_{\tau=j}^{j+p} [\lambda_i](v_{\tau-1},v_\tau) + \sum_{\tau=-q}^{0} [\lambda_i](w_\tau,w_{\tau+1})$,

(iii) $[\lambda_i](\bar{v}_{j-1},\bar{w}_1) = 0$ and $(\bar{v}_{j-1},\bar{w}_1)$ is an i-discontinuity in (u_ℓ,u_*),

(iv) $\displaystyle \sum_{\tau=1}^{j-1} |[\lambda_i](v_{\tau-1},v_\tau) - [\lambda_i](v_{\tau-1},v_\tau)|$

$\displaystyle \qquad\qquad + \sum_{\tau=2}^{k} |[\lambda_i](w_{\tau-1},w_\tau) - [\lambda_i](\bar{w}_{\tau-1},\bar{w}_\tau)| = O(1)\,\theta\alpha\beta$,

(v) $\displaystyle \sum_{\tau=1}^{j} \|(v_\tau-v_{\tau-1}) - (\bar{v}_\tau-\bar{v}_{\tau-1})\|$

$\displaystyle \qquad\qquad + \sum_{\tau=1}^{k} \|(w_\tau-w_{\tau-1}) - (\bar{w}_\tau-\bar{w}_{\tau-1})\| = O(1)\,\theta\alpha\beta$,

(vi) $(\alpha+\beta)\,\theta_\ell(u_\ell,u_*)$

$\qquad \leq \alpha\theta_\ell(u_\ell,u_m) + \beta[\theta_\ell(u_m,u_r) + \theta_\ell(u_\ell,u_m) + \theta_r(u_\ell,u_m)] + O(1)\,\theta\alpha\beta$,

$\qquad (\alpha+\beta)\,\theta_r(u_\ell,u_*) \leq \beta\theta_r(u_m,u_r)$

$\qquad\qquad\qquad + \alpha[(\theta_r(u_\ell,u_m) + \theta_\ell(u_m,u_r) + \theta_r(u_m,u_r)] + O(1)\,\theta\alpha\beta$,

(vii) $\theta_\ell(u_\ell,u_*) \leq \theta_\ell(u_\ell,u_m) + O(1)\,\theta\alpha\beta$,

$\qquad \theta_r(u_m,u_r) \leq \theta_r(u_\ell,u_*) + O(1)\,\theta\alpha\beta$.

PROOF: We will use Lemmas 5.1 and 5.2 repeatedly in this proof. When $\theta = 0$, the i-waves relating u_ℓ and u_r are the linear supperposition of those relating u_ℓ and u_m, and those relating u_m and u_r. In this case the lemma holds trivially. Thus we suppose that $\theta > 0$, and for definiteness $\theta_\ell(u_m,u_r) \geq \theta_r(u_\ell,u_m)$. Let (u_m,u_+) be the first i-discontinuity from the left in (u_m,u_r) so that $\theta_\ell(u_m,u_r) = \theta_\ell(u_m,v_+)$. We may use Lemmas 5.1 and 5.2 to show that as the result of the interaction of (u_m,u_+) with the elementary waves in (u_ℓ,u_m), there exists $\bar{u}_+$ on $T_i(u_\ell)$ with $\bar{u}_+^i = u_+^i$ so that u_ℓ is related to $\bar{u}_+$ by i-waves in (u_ℓ,u') and an i-discontinuity $(u',\bar{u}_+)$ for some u' on $T_i(u_\ell)$ between u_ℓ and u_m. Moreover, $\bar{u}_+$ satisfies

(7.1) $\|\bar{u}_+ - u_+\| = O(1)\,\theta\alpha_1\beta_1$, $\alpha_1 \equiv \|u' - u_m\|$, $\beta_1 \equiv \|u_m - u_+\|$,

(7.2) $(\alpha_1 + \beta_1) \sigma(u', \overline{u}_+) = \sum_j \alpha_i^j \hat{\lambda}_i^j + \beta_1 \sigma(u_m, u_+) + O(1) \theta \alpha_1 \beta_1$,

where α_i^j is the strength of an i-wave in $(u', \overline{u}_+)$ and $\hat{\lambda}_i^j$ the mean
value of λ_i of these waves. When $u_+ \neq u_r$, we choose $\overline{u}_r$ on $T_i(\overline{u}_+)$
with $\overline{u}_r^i = u_r$ and denote by θ the angle between $(u_\ell, \overline{u}_+)$ and $(\overline{u}_+, \overline{u}_r)$.
Note that Lemma 6.3 implies that the wave pattern in (u_+, u_r) is close to
that in $(\overline{u}_+, \overline{u}_r)$, and from (7.1) and (7.2), θ_1 is strictly less than θ.
Our next step is to allow the i-discontinuity $(u', \widetilde{u}_+)$ in $(u_\ell, \widetilde{u}_+)$ to
interact with $(\overline{u}_+, \overline{u}_r)$. Suppose that $\overline{\overline{u}}_r \in T_i(u')$ and θ_2 is the angle
between (u_ℓ, u') and $(u', \overline{\overline{u}}_r)$. Then analogous estimates as above also
hold for this interaction and θ_2 is strictly less than θ_1. Thus it is
clear that we may apply Lemma 6.3 to show that wave patterns are preserved
module certain error terms by reducing the general interaction of i-waves
into elementary interactions discussed in Lemmas 5.1 and 5.2. That the
sequence θ_n, $n = 1, 2, \ldots$, converge to zero and the total sum of error
terms do not exceed $O(1) \theta \alpha \beta$ can be verified using similar arguments used
in the proof of Lemma 5.2; details are omitted. Estimates (vi) and (vii)
are proved using estimates analogous to (7.1) and (7.2); details are also
omitted.   Q.E.D.

LEMMA 7.2. Suppose that $u_m \in T_i(u_\ell)$, $u_r \in T_i(u_m)$ and $u_\ell^i < u_r^i < u_m^i$.
Then there exists $\overline{u} \in T_i(u_\ell)$ with $\overline{u}^i = u_r^i$ such that

(i) $\|\overline{u} - u_r\| = O(1) \alpha \beta$, $\alpha \equiv \|u_\ell - u_r\|$, $\beta \equiv \|u_\ell - u_m\|$.

Moreover, there exist a partition $\{v_0, v_1, \ldots, v_{k_1}, \ldots, v_{k_2}, \ldots, v_{k_3}\}$ for
(u_ℓ, u_r), $0 < k_1 \leq k_2 \leq k_3$ and a partition $\{\overline{v}_0, \overline{v}_1, \ldots, \overline{v}_{k_1}, \ldots, \overline{v}_{k_2}\}$ for
$(u_\ell, \overline{u})$ with the property that $v_{k_2}^i = \overline{v}_{k_2}^i$, (v_{k_1}, v_{k_2}) is a part of an
i-discontinuity in (u_ℓ, u_m) and

(ii) $\displaystyle\sum_{j=k_1}^{k_2} [\lambda_i] (\overline{v}_{j-1} \overline{v}_j) \equiv P_{\lambda_i}(u_\ell, u_m, u_r) = O(1) C_i(u_\ell, u_m, u_r)$,

(iii) $\displaystyle\sum_{j=1}^{k_2} \| (v_j - v_{j-1}) - (\overline{v}_j - \overline{v}_{j-1}) \| = O(1) \alpha \beta$,

(iv) $\displaystyle\sum_{j=1}^{k_1 - 1} | [\lambda_i] (v_{j-1}, v_j) - [\lambda_i] (\overline{v}_{j-1}, \overline{v}_j) | = O(1) \alpha \beta$,

(v) $|\theta_\ell(u_\ell, u_r) - \theta_\ell(u_\ell, u_m)| + |\theta_r(u_\ell, u_r) - \theta_r(u_\ell, u_m)|$

     $= O(1) C_1(u_\ell, u_m, u_r)$,

where $C_i(u_\ell, u_m, u_r) \equiv \beta$ denotes the <u>amount of cancellation</u> of the strength of i-waves, P_{λ_i} is the <u>amount of production</u> of i-expansion waves, and the bounds $O(1)$, as in the last lemma, is dominated by $\max|\nabla \lambda_i \cdot r_i|$.

PROOF: From the construction of the curves T_i, it can be shown that T_i is tangent at each point to either an S_i or an R_i curve. Since an S_i curve bifurcates from R_i curve, Lemma 2.2, we see that

$$u_m \in R_i(\overline{u}) + O(1)\alpha^2\beta$$

$$u_r \in R_i(u_m) + O(1)\beta^3 .$$

Thus estimate (i) follows from that R_i curves are integral curves of the vector field r_i and that $\alpha > \beta$. Note that though (v_{k_1}, v_{k_2}) is part of an i-discontinuity in (u_ℓ, u_m), $(\overline{v}_{k-1}, \overline{v}_k)$ may contain i-rarefaction waves. Thus rarefaction waves may be produced due to cancellation of the strength of i-waves. The amount of expansion produced is denoted by P_{λ_i}. Estimate (ii) is proved by continuity arguments (c.f. Lemma 5.2); and (iii), (iv) follows from (i) and Lemma 6.3; details are omitted.   Q.E.D.

<u>THEOREM</u> 7.3. Suppose that the Riemann problems (u_ℓ, u_m), (u_m, u_r) and (u_ℓ, u_r) are solved by i-waves (v_{i-1}, v_i), (w_{i-1}, w_i) and (u_{i-1}, u_i), $i = 1, 2, \ldots, n$, respectively and all waves under consideration are weak. Then

(i) $\quad u_i = v_i + \displaystyle\sum_{k=1}^{i} (w_k - w_{k-1}) + O_1(1)Q_s(u_\ell, u_m, u_r) + O_2(1)Q_d(u_\ell, u_1, u_2)$

(ii) $\quad u_i = w_i + \displaystyle\sum_{k=i+1}^{n} (v_k - v_{k-1}) + O_1(1)Q_s(u_\ell, u_m, u_r) + O_2(1)Q_d(u_\ell, u_m, u_r)$

where Q_s and Q_d measure the effect of interactions of waves of the same family and of different families respectively:

$$Q_s(u_\ell, u_m, u_r) \equiv \sum_{i=1}^{n} Q_s^i(u_\ell, u_m, u_r) ,$$

$$Q_s^i(u_\ell, u_m, u_r) \equiv \theta_i \alpha_i \beta_i \quad \text{if} \quad (v_i^i - v_{i-1}^i)(w_i^i - w_{i-1}^i) \geq 0 ,$$

$$\theta_i \equiv \theta_r(v_{i-1}, v_i) + \theta_\ell(w_{i-1}, w_i) ,$$

$$\alpha_i \equiv \|v_{i-1} - v_i\|, \qquad \beta_i \equiv \|w_{i-1} - w_i\|,$$

$$Q_s^i(u_\ell, u_m, u_r) \equiv \alpha_i \beta_i \quad \text{if} \quad (v_i^i - v_{i-1}^i)(w_i^i - w_{i-1}^i) < 0,$$

$$Q_d(u_\ell, u_m, u_r) = \sum_{i > j} \alpha_i \beta_j.$$

Here the bounded $O_1(1)$ is dominated by $\max|\nabla\lambda_i \cdot r_i|$ and $Q_2(1)$ by $\max\|f''\|$.

The above theorem relates the strength of waves before and after interaction. The change in wave patterns due to interactions is studied in the following theorem. The strength of i-waves is now measured by u^i:
$\alpha_i \equiv |v_{i-1}^i - v_i^i|$, etc.

THEOREM 7.4. Assume that the hypothesis of Theorem 7.3 holds, then there exist partitions $\{\bar{v}_{-j_0}, \ldots, \bar{v}_0, \bar{v}_1, \ldots, \bar{v}_j, \ldots, \bar{v}_{j_1+j_2}\}$ for (v_{i-1}, v_i), $\{\bar{w}_{-k_0}, \ldots, \bar{w}_0, \bar{w}_1, \ldots, \bar{w}_{k_1}, \ldots \bar{w}_{k_1+k_2}\}$ for (w_{i-1}, w_i), and $\{\bar{u}_{-\ell_0}, \ldots, \bar{u}_0, \bar{u}_1, \ldots, \bar{u}_{\ell_1}, \ldots, \bar{u}_{\ell_1+\ell_2}\}$ for (u_{i-1}, u_i) with the following properties: Suppose that $(v_{i-1}^i, v_{i-1}^i)(w_{i-1}^i - w_i^i) \geq 0$; for definiteness, we assume $v_i^i > v_{i-1}^i$ and $w_i^i > w_{i-1}^i$. Then $\ell_1 = j_1 + k_1$, $[\lambda_i](\bar{u}_{\ell_1-1}, \bar{u}_{\ell_1}) = 0$ and there exists $C_{\lambda_i} \equiv C_{\lambda_i}(u_\ell, u_m, u_r) \geq 0$ such that

(i) $\displaystyle\sum_{j=j_1}^{j_1+j_2} [\lambda_i](\bar{v}_{j-1}, \bar{v}_j) + \sum_{k=-k_0}^{0} [\lambda_i](\bar{w}_k, \bar{w}_{k+1}) = C_{\lambda_i} + O(1)Q(u_\ell, u_m, u_r)$,

(ii) $\displaystyle\sum_{j=-j_s}^{0} \|\bar{v}_j - \bar{v}_{j+1}\| + \sum_{k=k_1}^{k_1+k_2} \|\bar{w}_{k-1} - \bar{w}_k\| + \sum_{\ell=-\ell_0}^{0} \|\bar{u}_\ell - \bar{u}_{\ell+1}\| + \sum_{\ell=\ell_1}^{\ell_1+\ell_2} \|\bar{u}_{\ell-1} - \bar{u}_\ell\|$

$= O(1)Q(u_\ell, u_m, u_r)$,

(iii) $\displaystyle\sum_{\tau=2}^{j_1-1} \|(\bar{v}_\tau - \bar{v}_{\tau-1}) - (\bar{u}_\tau - \bar{u}_{\tau-1})\| + \sum_{\tau=2}^{k_1-1} \|(\bar{w}_\tau - \bar{w}_{\tau-1}) - (\bar{u}_{\tau-j_1} - \bar{u}_{\tau-j_1-1})\|$

$= O(1)Q(u_\ell, u_m, u_r)$,

(iv) $\displaystyle\sum_{\tau=2}^{j_1-1} |[\lambda_i](\bar{v}_{\tau-1}, \bar{v}_\tau) - [\lambda_i](\bar{u}_{\tau-1}, \bar{u}_\tau)|$

$+ \displaystyle\sum_{\tau=2}^{k_1-1} |[\lambda_i](\bar{w}_\tau - \bar{w}_{\tau-1}) - [\lambda_i](\bar{u}_{\tau-j} - \bar{u}_{\tau-j-1})| = O(1)Q(u_\ell, u_m, u_r)$,

(v) $(\alpha_i+\beta_i)\theta_\ell(u_{i-1},u_i) \leq \alpha_i\theta_\ell(v_{i-1},v_i) + \beta_i[\theta_\ell(w_{i-1},w_i) + \theta_\ell(v_{i-1},v_i)$

$+ \theta_r(v_{i-1},v_i)] + O_1(1)Q_s(u_\ell,u_m,u_r) + O_2(1)(\alpha_i+\beta_i)Q_d(u_\ell,u_m,u_r),$

$(\alpha_i+\beta_i)\theta_r(u_{i-1},u_i) \leq \beta_i\theta_r(w_{i-1},w_i) + \alpha_i[\theta_r(v_{i-1}v_i) + \theta_\ell(w_{i-1},w_i)$

$+ \theta_r(w_{i-1},w_i)] + O_1(1)Q_s(u_\ell,u_m,u_r) + O_2(1)(\alpha_i+\beta_i)Q_d(u_\ell,u_m,u_r),$

$$\alpha_i \equiv |v_i^i - v_{i-1}^i|, \qquad \beta_i \equiv |w_i^i - w_{i-1}^i|.$$

(vi) $\theta_\ell(u_{i-1},u_i) \leq \theta_\ell(v_{i-1},v_i) + O(1)Q(u_\ell,u_m,u_r)$

$\theta_r(w_{i-1},w_i) \leq \theta_r(u_{i-1},u_i) + O(1)Q(u_\ell,u_m,u_r).$

Suppose that $(v_{i-1}^i-v_i^i)(w_{i-1}^i-w_i^i) < 0$, for definiteness, we assume that $v_i^i - v_{i-1}^i \geq w_{i-1}^i - w_i^i > 0$. Then $k_0 = 0$, $j_1 = \ell_1$, $j_2 = \ell_2+k_1+k_2$ and there exist $P_{\lambda_i} \equiv P_{\lambda_i}(u_\ell,u_m,u_r) \geq 0$ such that

(vii) $\displaystyle\sum_{\ell=\ell_1}^{\ell_1+\ell_2} [\lambda_i](\bar{u}_\ell,\bar{u}_{\ell+1}) = P_{\lambda_i} = O(1)c_i,$

(viii) $\displaystyle\sum_{j=-j_0}^{0} \|\bar{v}_j - \bar{v}_{j+1}\| + \sum_{k=k_1}^{k_1+k_2} \|\bar{w}_k - \bar{w}_{k+1}\| + \sum_{j=j_1+j_2-k_2}^{j_1+j_2} \|\bar{v}_{j-1} - \bar{v}_j\|$

$\displaystyle + \sum_{\ell=-\ell_0}^{0} \|\bar{u}_\ell - \bar{u}_{\ell+1}\| = O(1)C_i + O(1)Q(u_\ell,u_m,u_r),$

$$C_i \equiv \min(|\alpha_i|,|\beta_i|),$$

(ix) $\displaystyle\sum_{\tau=1}^{\ell_1-1} \|(\bar{v}_\tau-\bar{v}_{\tau-1}) - (\bar{u}_\tau-\bar{u}_{\tau-1})\| = O(1)Q(u_\ell,u_m,u_r),$

(x) $\displaystyle\sum_{\tau=1}^{\ell_1-1} |[\lambda_i](\bar{v}_{\tau-1},\bar{v}_\tau) - [\lambda_i](\bar{u}_{\tau-1},\bar{u}_\tau)| = O(1)Q(u_\ell,u_m,u_r),$

(xi) $|\theta_\ell(u_{i-1},u_i) - \theta_\ell(v_{i-1},v_i)| + |\theta_r(u_{i-1},u_i) - \theta_r(v_{i-1},v_i)|$

$$= O(1)Q(u_\ell,u_m,u_r).$$

PROOF OF THEOREMS 7.3 AND 7.4: Denote by α_i, β_i and γ_i, $i = 1,2,\ldots,n$, the strengths of i-waves in (u_ℓ,u_m), (u_m,u_r) and (u_ℓ,u_r) respectively and define $\phi_i(\alpha,\beta) \equiv \gamma_i - \alpha_i - \beta_i$. It is clear from the smoothness and stableness of T_i curves that f_i is second order smooth. When $\alpha_{k+1} = \cdots = \alpha_n = 0$ and $\beta_1 = \beta_2 = \cdots = \beta_k = 0$ for some k it is clear that the solution for (u_ℓ,u_r) is the linear superposition of those for (u_ℓ,u_m) and (u_m,u_r) and thus $\phi_i(\alpha,\beta) = 0$. As a first step, we deal with the case when $Q_s^i = 0$ for all $i = 1,2,\ldots,n$. Thus we assume that $\alpha_1 = \alpha_2 = \cdots = \alpha_k = 0$ and $\beta_{k+1} = \cdots = \beta_n = 0$ for some k. In this case, as just noted, $\phi_i(\alpha,0) = \phi_i(0,\beta) = 0$ and so by the smoothness of ϕ_i, we conclude that $\phi_i(\alpha,\beta) = O(1)|\alpha||\beta| = O(1)Qd(u_\ell,u_m,u_r)$

(7.3) $\gamma_i = \alpha_i + \beta_i + O(1)Q(u_\ell,u_m,u_r).$

Next we establish the above estimate when $\alpha_i = 0$ for all $i \neq k$ and β_i, $i = 1,2,\ldots,n$, are arbitrary. This is done in three steps. In the first step, $(u_\ell,u_m) \equiv (v_{k-1},v_k)$ interacts with (u_m,w_{k-1}). We have just studied this interaction and showed that the i-waves, $i > k$, in (u_ℓ,w_{k-1}) is of strength $O(1)Q(u_\ell,v_{k-1},w_{k-1}) \leq O(1)Q(u_\ell,u_m,u_r)$, and the strength of the k-wave $(\bar{u}_{k-1},\bar{u}_k)$ in (u_ℓ,w_{k-1}) differs from α_k also by $O(1)Q(u_\ell,u_m,u_r)$. Next we let $(\bar{u}_{k-1},w_{k-1})$ interact with (w_{k-1},w_k). Although we have interaction of more than two k-waves, since other waves have small strength, we may use the continuity arguments and Lemmas 7.1 and 7.2 to treat this interaction. Finally, we note that (u_ℓ,w_k) and (w_k,u_r) are noninteracting module an error of the order of $Q(u_\ell,u_m,u_r)$. Thus this final interaction is handled easily by continuity arguments.

To establish (7.3) for general interactions, we use induction process. We first let (v_{n-1},v_n) interact with (u_m,u_r) which has just been studied. Suppose that we have proved (7.3) for the interaction of (v_k,u_m) and (u_m,u_r). Again we use the above result to study the interaction of (v_{k-1},v_k) and (v_k,u_r). This proves (7.3) for general interactions; details are omitted.

We note that in the above arguments, we may use any nonsingular parameter along T_i curves to measure the strength of waves. Thus (7.3) implies:

(7.4) $u_i - u_{i-1} = (v_i-v_{i-1}) + (w_i-w_{i-1}) + O(1)Q(u_\ell,u_m,u_r),$

whence Theorem 7.3 follows easily. The estimates on the changes in wave patterns follow from similar arguments using Lemmas 6.3, 7.1 and 7.2; details are omitted.   Q.E.D.

§8. Nonlinear functional

We first describe briefly the Glimm scheme, [15], [31]. Choose an equidistributed sequence $\{a_k\}_{k=1}^{\infty}$ in $(-1,1)$ and mesh length $r = \Delta x$, $s = \Delta t$ satisfying Courant-Friedrichs-Lewy condition:

$$(C\text{-}F\text{-}L) \qquad \frac{r}{s} \geq \max_{i=1,\ldots,n} \{\lambda_i(u)\}$$

over all u under consideration. An approximate solution $u_r(x,t) \equiv u_r(x,t;a_k)$ for (1.1) and (1.2) is defined inductively as follows: Suppose that $u_r(x,t)$ has been defined for $0 \leq t < ks$. Then we set

$$u_r(x,ks) = u_r((h+a_k)r,ks-0) \quad \text{for} \quad (h-1)\ell < x < (h+1)r, \qquad h + k = \text{odd}.$$

Thus $u_r(x,ks)$ is a step function. By resolving the discontinuities at $x = hr$, $k + h = $ even, $u_r(x,t)$ is defined for $ks \leq t < (k+1)s$ and consists of elementary waves (see Section 4). These waves do not interact in the layer $ks < t < (k+1)s$ because of the (C-F-L) condition. A space-like curve J is called on I-wave if J is piecewise linear and passes through mesh points of the form $A_{h,k} = ((h+a_k)r,ks)$, $h + k = $ odd. A diamond $\Delta = \Delta_{n,k}$, $h + k = $ even, is a polygon with vertices $A_{n-1,k}$, $A_{n+1,k}$, $A_{n,k+1}$ and $A_{h,k-1}$. An I-curve J_2 is an immediate successor of J_1 if J_2 and J_1 sandwich a diamond and J_2 lies toward larger time than J_1. The I-curve between $t = 0$ and $t = s$ is denoted by O.

To obtain estimates for approximate solutions, we define the following nonlinear functional $F(J)$ for each I-curve J:

$$F(J) \equiv L(J) + KQ_s(J) + KQ_d(J)$$

$$L(J) \equiv \sum_{i=1}^{n} L_i(J) \equiv \sum_{i=1}^{n} \{|\alpha|: \alpha \text{ is the strength of any } i\text{-wave crossing } J\}$$

$$Q_d(J) \equiv \sum\{|\alpha\beta|: \alpha(\beta) \text{ is the strength of any } i\text{-wave } (j\text{-wave}) \text{ crossing } J, \; i > j, \text{ and } \alpha \text{ lies to the left of } \beta\}$$

$$Q_s(J) \equiv \sum_{i=1}^{n} Q_s^i(J)$$

$$Q_s^i(J) \equiv \sum\{\phi(\alpha,\beta): \alpha \text{ and } \beta \text{ are strengths of } i\text{-waves } (u_-,u_+) \text{ and } (\bar{u}_-,\bar{u}_+) \text{ respectively crossing } J \text{ and } (u_-,u_+) \text{ lies to the left of } (\bar{u}_-,\bar{u}_+)\},$$

$$\phi(\alpha,\beta) \equiv \begin{cases} \theta(u_-,u_+;\overline{u}_-,\overline{u}_+)\,|\alpha\beta| & \text{if} \quad (u_-^i-u_+^i)(\overline{u}_-^i-\overline{u}_+^i) \geq 0 \\[2ex] \max|\nabla\lambda_i r|\,|\alpha\beta| & \text{if} \quad (u_-^i-u_+^i)(\overline{u}_-^i-\overline{u}_+^i) \leq 0 \end{cases}$$

$$\theta(u_-,u_+;\overline{u}_-,\overline{u}_+) \equiv \theta_r(u_-,u_+) + \theta_\ell(\overline{u}_-,\overline{u}_+) + \sum\{\theta_\ell(u_1,u_2) + \theta_r(u_1,u_2): (u_1,u_2)$$

is any i-wave crossing J between (u_-,u_+) and $(\overline{u}_-,\overline{u}_+)\}$

Remark: As in [15], the functional is so designed that it is a nonincreasing function of time. Notice that the potential interaction of waves of the same family and same direction is defined in terms of wave strengths and also of the angle between the waves. A simple heuristic explanation for this definition can be given as follows: Suppose that the waves crossing an I-curve J_1 consist of i-discontinuities only. Assume that two i-discontinuities (u_0,u_1) and (u_1,u_2) interact in a diamond Δ and J_2 is the immediate successor of J_1 with respect to Δ. Let us define the amount of interaction in Δ by

$$Q(\Delta) \equiv \theta\alpha_1\alpha_2$$

where $\theta = \theta_r(u_0,u_1) + \theta_\ell(u_1,u_2) = \sigma(u_0,u_1) - \sigma(u_1,u_2)$ is the angle between these two waves and $\alpha_1 = |u_0^i - u_1^i|$, $\alpha_2 = |u_1^i-u_2^i|$ are their strengths. In view of (iii) of Lemma 5.1, $Q(\Delta)$ does measure the effect of nonlinear interaction of these waves. Now, for simplicity, we <u>suppress</u> this effect and assume that the interaction produces simply another i-discontinuity (u_0,u_2). When this was the case $L(J_1) = L(J_2)$. Let us compare $Q(J_1)$ and $Q(J_2)$. The speed of (u_0,u_2) can be computed from (iii) of Lemma 5.1:

$$\sigma(u_0,u_2) = \frac{\alpha_1}{\alpha_1+\alpha_2}\sigma(u_0,u_1) + \frac{\alpha_2}{\alpha_1+\alpha_2}\sigma(u_1,u_2).$$

Let (u_-,u_+) be any i-discontinuity to the right of (u_0,u_2) and crossing $J_1 \cap J_2$ with strength $\gamma = |u_-^i - u_+^i|$. We have:

$$\theta(u_0,u_1;u_-,u_+) = \theta_r(u_0,u_1) + \theta_\ell(u_-,u_+) + (\theta_\ell(u_1,u_2) + \theta_r(u_1,u_2)) + \overline{\theta}$$

$$\theta(u_1,u_2;u_-,u_+) = \theta_r(u_1,u_2) + \theta_\ell(u_-,u_+) + \overline{\theta}$$

$$\theta(u_0,u_2;u_-,u_+) = \theta_r(u_0,u_2) + \theta_\ell(u_-,u_+) + \overline{\theta}$$

Detailed readings of the definition of $Q(J)$ and the estimates in Theorems 7.3 and 7.4 show that

(8.1) $Q(J_2) - Q(J_1) \leq [-1 + O(1)L(J_1)]Q(\Delta)$.

The proof of this estimate, rather lengthly but entirely straightforward, is left to the reader (c.f. [15], [30] and the above remark). Finally, it follows from the last three estimates that

$$F(J_2) - F(J_1) \leq Q(\Delta)[O(1) - K + O(1)K \text{ T.V.}]$$

which is nonpositive if K is sufficiently large and $O(1)\text{T.V.}$ sufficiently small.   Q.E.D.

   Given a region $\wedge$ composed of diamonds, we set

$$Q_s(\wedge) \equiv \sum_{\Delta \in \wedge} Q_s(\Delta), \qquad Q_d(\wedge) \equiv \sum_{\Delta \in \wedge} Q_d(\Delta),$$

$$Q(\wedge) \equiv Q_s(\wedge) + Q_d(\wedge).$$

When the i-waves entering a diamond Δ are of the same direction (c.f. (i)$\sim$(vi) of Theorem 7.4), the combining of i-waves may eliminate some i-expansion waves. In consistency with the notation used in Theorem 7.4, the amount of the expansion of such i-wave is denoted by $C_{\lambda_i}(\Delta)$. When the i-waves entering Δ are of opposite direction (c.f. (vii)$\sim$(xii) of Theorem 7.4); the amount of increase of the expansion of i-wave due to cancelling of i-waves is denoted by $P_{\lambda_i}(\Delta)$. Similarly we define $C_{\lambda_i}(\wedge)$, $P_{\lambda_i}(\wedge)$ and the amount of cancelling of i-waves $C(\wedge)$ in $\wedge$.

COROLLARY 8.2. Suppose that J is an I-curve which lies toward smaller time than the region $\wedge$. Then

   (i) $Q(\wedge) \leq 2Q(J)$,

   (ii) $C_i(\wedge) \leq L_i(J) + O(1)Q(J)$, $i = 1,2,\ldots,n$,

   (iii) $P_{\lambda_i}(\wedge) = O(1)[C_i(\wedge) + Q(J)]$,

   (iv) $C_{\lambda_i}(\wedge) \leq L_{\lambda_i}(J) + O(1)[P_{\lambda_i}(\wedge) + Q(J)]$,

where $L_{\lambda_i}(J)$ is the total amount of expansion of i-waves crossing J:

$$L_{\lambda_i}(J) \equiv \sum\{[\lambda_i](u_-,u_+), \quad (u_-,u_+) \text{ any i-wave crossing } J\}.$$

PROOF. It follows from the estimate (8.1) that for small T.V.,

$$Q(\Delta) \leq 2[Q(J_1) - Q(J_2)].$$

By summing up the above inequality over all Δ in $\wedge$ we obtain (i). It is clear that the total amount of cancellation of i-waves in $\wedge$, $C_i(\wedge)$, can never exceed the sum of the total amount of i-waves that were there before interaction, which is $L_i(J)$, and the amount of i-waves produced due to interactions, which is $O(1) \ Q(\wedge)$. Thus (ii) follows from (i). We have noticed in Section 7 that when expansion i-waves are produced due to cancelling of i-waves, the amount of this production is dominated by the amount of waves cancelled (c.f. (viii), Theorem 7.4). Thus we have

$$P_{\lambda_i}(\wedge) = O(1) \ [C_i(\wedge) + Q(\wedge)$$

and (iii) follows from (i) and (ii). Finally, the amount of cancellation of expansion waves is clearly no larger than the sum of the amount of expansion waves before interaction, which is $L_{\lambda_i}(J)$, the amount produced due to the cancellation of i-waves, which is $P_{\lambda_i}(\wedge)$, and the amount created due to interactions, which is dominated by $Q(\wedge)$. Consequently, (iv) follows from (i) and (iii).   Q.E.D.

§9. Wave partition

We will partition the waves in the approximate solution $u_r(x,t) \equiv u_r(x,t;a_k)$ in the region $\wedge \equiv \{(x,t): N_1 s \leq t \leq N_2 s\}$. The elementary i-waves issued from (hr,ks), $h + k$ = even, are partitioned by $\{v_{j(h,k)}\}$ according to Definition 6.1, and as before $[\lambda_i]_{j(h,k)} = \lambda_i(v_{j(h,k)}) - \lambda_i(v_{j(h,k)-1}) > 0$ if $(v_{j(h,k)-1}, v_{j(h,k)})$ is an i-rarefaction wave, and $[\lambda_i]_{j(h,k)} = 0$ if $(v_{j(h,k)-1}, v_{j(h,k)})$ is part of an i-discontinuity (u_-, u_+). Similarly, the "speed" $\sigma_{i,j(h,k)} \ (v_{j(h,k)-1}, v_{j(h,k)})$ is defined to be $\lambda_i(v_{j(h,k)})$ or $\sigma(u_-, u_+)$ respectively.

THEOREM 9.1. Given any $\varepsilon > 0$ there exist partitions $\{v_{j(h,k)}\}$, $h + k$ = even, $(h,k) \in \wedge$, with the following properties: For each fixed (h,k), $\{j(h,k)\}$ can be written as a disjoint union of $I(h,k)$, $II(h,k)$, $III(h,k)$ and $IV(h,k)$, $I(k) \equiv \bigcup_{h=-\infty}^{\infty} I(h,k)$, etc., so that for any k, $N_1 \leq k \leq N_2$.

   (i) $a_k \notin (\sigma_{i,j(h,k)}, \sigma_{i,j(h,k)-1})$,

   (ii) $\sum \{ \|v_{j(h,k)} - v_{j(h,k)-1}\| : j(h,k) \in IV(k) \} = O(1) \ [Q(\wedge) + C_i(\wedge)],$

and there is an one-to-one corresponding between $I(k)$ and $I(N_1)$, between $II(k)$ and $II(N_i)$, etc., denoted by

$$(h = h(h_0,k,j_0),\ j = j(h_0,k,j_0)) \longleftrightarrow j_0(h_0,N_1)$$

so that for any $(h_0,j_0) \in I(N_1) \cup II(N_1) \cup III(N_1) \cup IV(N_1)$,

    (iii)   (monotonicity) either $h(h_0,k,j_0) > h(h_1,k,j_1)$, or $h(h_0,k,j_0)$
            $= h(h_1,k,j_1)$ and $j(h_0,k,j_0) > j(h_1,k,j_1)$ if either $h_0 > h_1$
            or $h_0 = h_1$ and $j_0 > j_1$,

    (iv)   (determinancy) $h(h_0,k+1,j_0) = h(h_0,k,j_0) \pm 1$ if
        $s\lambda_{i,j(h(h_0,k,j_0),k)} - ra_k \gtreqless 0$,

    (v) $\sum\{ \max_{N_1 \le k \le N_2} \| (v_{j(h_0,k,j_0)} - v_{j(h_0,k,j_0)-1}) - (v_{j_0(h_0,N_1)} - v_{j_0(h_0,N_1)-1}) \|$
        $: j_0(h_0,N_1) \in I(N_1) \cup II(N_1) \cup III(N_1)\} = O(1)Q(\wedge)$,

    (vi) $\sum\{ \max_{N_1 \le k \le N_2} |\lambda_{i,j_0(h_0,N_1)} - \lambda_{i,j(h_0,k,j_0)}| \| v_{j_0(h_0,N_1)} - v_{j_0(h_0,N_1)-1}\|$
        $: j_0(h,N_1) \in I(N_1) \cup II(N_1) \cup III(N_1)\} = O(1)Q(\wedge)$,

    (vii) $[\lambda_i]_{j_0(h_0,k,j_0)} < \epsilon$, $j_0(h_0,N_1) \in I(N_1) \cup II(N_1) \cup II(N_1)$.

For $j_0(h_0,N_1) \in I(N_1)$, $(v_{j(h_0,k,j_0)-1}, v_{j(h_0,k,j_0)})$ is an i-rarefaction wave and

    (viii) $\sum\{|[\lambda_i]_{j(h_0,k,j_0)} - [\lambda_i]_{j_0(h_0,N_1)}| : j_0(h_0,N_1) \in I(N_1)\} = O(1)Q(\wedge)$;

for $j_0(h_0,N_1) \in II(N_1)$, $(v_{j(h_0,k,j_0)-1}, v_{j(h_0,k,j_0)})$ is part of an i-discontinuity, and for $j_0(h_0,N_1) \in III(N_1)$, $\{(v_{j(h_0,k,j_0)-1}, v_{j(h_0,k,j_0)}):$ $N_1 \le k \le N_2\}$ consists of both i-rarefaction waves and parts of i-discontinuities, and for each fixed k between N_1 and N_2

    (ix) $\sum\{[\lambda_i]_{j(h_0,k,j_0)} : j_0(h_0,N_1) \in III(N_1)\} \le C_{\lambda_i}(\wedge) + P_{\lambda_i}(\wedge)$.

PROOF: The proof is based on the local partitioning results stated in Theorems 7.3 and 7.4. We first note that in these theorems when the

partition of incoming waves (or outgoing waves) is refined then outgoing
waves (or incoming waves) can be further partitioned accordingly so that
the estimates still hold. The proof of the theorem is by induction. Let
J_1 and J_2 be any two I-curves, J_2 an immediate successor of J_1, and
Δ be the diamond between them. Suppose that waves crossing and below J_1
have already partitioned. By Theorem 7.3, 7.4 and the above remark, we may
refine the partition of the waves entering Δ and partition the waves
leaving Δ so that the estimates in Theorem 7.4 hold. Notice that the
waves entering Δ have already been partitioned so as to relate to the
waves beneath J_1 by induction hypothesis. Now that we are refining the
partition of the waves entering Δ, we must also refine the partition of
those waves below J_1 which are related to the waves entering Δ. As in
the case with partitions in Theorem 7.4, such refinements can always be
done by inductional process. We now indicate how the estimates in the
theorem are verified. First, estimates (i) and (vii) hold if the partitions are fine enough. (iii) and (iv) follow from the construction of
the scheme at the beginning of Section 8. (vi) follows from Theorems 7.3
and 7.4. The rest of the estimates follow from Theorem 7.4. This completes the proof of the theorem.   Q.E.D.

The above theorem provides a setting to investigate the expansion and
compression of waves, which will be done in Section 11. To this end, we
introduce (approximate) _generalized i-characteristics_, χ_i, $i = 1, 2, \ldots, n$,
of types I and II which are curves, move along the i-waves in the approximate solution $u_r(x, t; a_k) \equiv u_r(x, t)$ and passes through $h(h_0, k, j_0)$ for
each point (h_0, N_1) with $j_0(h_0, N_1)$ being in $I(N_1)$ and $II(N_2)$ respectively. For definiteness, when χ_i is of type I, it moves either always
along the right edge of the corresponding i-rarefaction waves or always
along the left edge of the i-rarefaction waves. Thus, χ_i is not a
Lipschitz continuous curve; it is time-like and jumps to the right or left
at $t = ks$, $N_1 \leq k \leq N_2$, according to the rule (iv) of the above theorem.
Characteristics of type I propagate with characteristic speed and those of
type II propagate with shock speed. One may also define characteristics
corresponding to $III(N_1)$ and $IV(N_1)$; however, we will not consider those
characteristics which are not needed in the present work. In contrast to
genuinely nonlinear system, [15], two characteristics may occupy same position at some time and divert from each other at latter time. Let $\Lambda_1 \subset \Lambda$
be the region between two i-characteristics χ_i and χ_1^2. The terms
$C_i(\Lambda_1)$, $P_{\lambda_i}(\Lambda_1)$, $Q(\Lambda_1)$, etc, can be defined in an obvious way. Let
$E_i(\Lambda_1)$ and $L_i(\Lambda_1)$ be the total strength of i-waves entering Λ_1 and
leaving Λ_1, respectively, through $t = N_1$ and $t = N_2$, and, $E_{\lambda_i}(\Lambda_1)$
and $L_{\lambda_i}(\Lambda_1)$ the total amount of expansion waves (as measured by $[\lambda_i]$,
the rate of expansion of these waves) entering and leaving Λ_1, respectively. Note that i-waves may cross χ_i^1 and χ_i^2 only due to interactions.
Thus the following lemma on the conservation of wave patterns follows from

the above theorem

LEMMA 9.2.

   (i) $L_i(\Lambda_1) = E_i(\Lambda_1) - C_i(\Lambda_1) + O(1)Q(\Lambda_1)$

   (ii) $L_{\lambda_i}(\Lambda_1) = E_{\lambda_i}(\Lambda_1) - C_{\lambda_i}(\Lambda_1) + P_{\lambda_i}(\Lambda_1) + O(1)Q(\Lambda_1).$

By the construction of χ_i, it is discontinuous at $t = ks$. We define
a Lipschitz continuous curve $\overline{\chi}_i$ which is parallel to χ_i for $(k-1)s <$
$t < ks$ for each integer k, and $\overline{\chi}_i$ passes through a same mesh point,
say $(h(h_0,k,j_0),k)$ for some k, as χ_i. Although χ_i and $\overline{\chi}_i$ do not
coincide, they approach each other as the mesh length r,s tend to zero
provided that the sequence is equidistributed.

Definition 9.3. A sequence $\{a_k\}_{k=1}^{\infty}$, $-1 < a_k < 1$, is equidistributed in
$(-1,1)$ if

$$\lim_{N \to \infty} A(N) = 0$$

$$A(N) \equiv \max_{I} \{\frac{A(N,I)}{N} - m(I): I \text{ any subinterval of } (-1,1)\}$$

where $A(N,I)$ denotes the number of k, $1 \le k \le N$, such that $a_k \in I$,
and $m(I)$ is the length of I.

LEMMA 9.4. (i) The speed of the curve $\overline{\chi}_i$ has total variation $O(1)$T.V.,
T.V. the total variation of the initial data.

   (ii) Suppose $T_2 = N_2 s$ and $T_1 = N_1 s$ are fixed, and $k^2 \le N_2 - N_1 \le$
         $(k+1)^2$. Then as $k \to \infty$, the distance between χ_i and $\overline{\chi}_i$ is
         $O(1)(T_2-T_1)[A(k) + k^{-1}\text{T.V.}].$

PROOF: The curve $\overline{\chi}_i$ is linear for $t \neq ks$, k any integer, and
changes speed only due to either wave interactions or j-waves, $j \neq i$
crossing χ_i. The total amount of j-waves crossing χ_i is less than the
total amount of j-waves crossing $t = N_1 s$ plus the total amount of wave
interactions in Λ. Thus (i) follows from Theorem 8.1 and Corollary 8.2.
We now derive (ii) from (i). For simplicity we assume that χ_i and $\overline{\chi}_i$
pass through the same mesh point on $t = N_1$. Denote by T.V.(j), $j =$
$1,2,\ldots,k$ the total variation of the speed of $\overline{\chi}_i$ for $(N_1+(j-1)k)s < t$
$\le (N_1+jk)s$. We have just shown that

$$\sum_{j=1}^{k} \text{T.V.}(j) = O(1)\text{T.V.}$$

and $O(1)$ is independent of k. We first investigate the distance D_1 between χ_i and $\bar{\chi}_i$ at $t = N_1 + ks$. Let χ_i^1 be a straight line with the same speed and position as χ_i and $\bar{\chi}_i$ at $t = N_1 s$. Since the speed of $\bar{\chi}_i$ for $t \in (N_1 s, N_1 s + ks)$ lies in an interval I_1 of length T.V.(1), we have

$$\text{distance}(\bar{\chi}_i, \chi_i^1) \leq ks\,T.V.(1).$$

Moreover, when Definition 9.3 is applied to $I \equiv \{asr^{-1}: a \in I_1\} \subset (-1,1)$ and its complement in $(-1,1)$, it follows from (vi) of Theorem 9.1 that

$$\text{distance}(\chi_i, \chi_i^1) \;=\; 2[T.V.(1)sr^{-1} + A(k)]ks$$

$$=\; O(1)ks[A(k) + T.V.(1)].$$

The above arguments can be applied also to other zones. Thus we have from the above estimates that

$$\text{distance}(\chi_i, \bar{\chi}_i) \;=\; O(1)\;ks[kA(k) + \sum_{j=1}^{k} T.V.(j)]$$

which proves (ii) of the lemma.   Q.E.D.

§10. Convergence of approximate solutions

The L_1-convergence of approximate solutions has been obtained in [15] through stochastic scheme. Our purpose in this section is to establish the local convergence through deterministic scheme. More precise results will be presented in Sections 12~15 along with the study of the singularities of the solution. We begin by describing some elementary results on scalar functions in one variable with bounded variation. We know that a function $w(x)$ with bounded variation can be written as a sum of a nondecreasing function $w^+(x)$ and a nonincreasing function $w^-(x)$ such that their total variations are related by $T.V.w^+ + T.V.w^- = T.V.w$.

Suppose that $\{w_i = w_i^+ + w_i^-,\ i = 1,2,\ldots\}$ is a sequence of functions with uniform variation and oscillation $|w_i|_\infty + T.V.w_i \leq M$. For each fixed x, $\{w_i^+(x)\}$ and $\{w_i^-(x)\}$ are bounded sequences and so there exist subsequences $\{w_{ij}^+(x)\}$ and $\{w_{ij}^-(x)\}$ such that

$$\lim_{ij \to \infty} w_{ij}^+(x) \;\equiv\; w_\infty^+(x),$$

$$\lim_{ij \to \infty} w_{ij}^-(x) \;\equiv\; w_\infty^-(x)$$

exist. By diagonal process we may assume that the above holds for all rational x. For x not rational, we choose a sequence $\{x_i\}$ of rational numbers, $x_i \to x - 0$, and define

$$w_\infty^\pm(x) \equiv \lim_{x_1 \to x-0} w_\infty^\pm(x_i)$$

$$w_\infty(x) \equiv w_\infty^+(x) + w_\infty^-(x).$$

Since it is clear that $w^+(x)$ (or $w^-(x)$), x rational, is a nondecreasing (nonincreasing) function, the functions $w_\infty^\pm(x)$, x real, are well-defined and $|w_\infty|_\infty + T.V.w_\infty \leq M$. We first study the local convergence of $\{w_{ij}\}$ to w_∞.

<u>Definition</u> 10.1. A sequence of scalar functions $\{f_i(x), x \in \mathbb{R}^n\}$ converges to $f(x)$ <u>locally</u> at $x = x_0$ if $\lim_{i \to \infty} f_i(x_0) = f(x_0)$ and given any $\delta > 0$, there exist $I = I(\delta) > 0$ and a neighborhood $N(\delta,x_0)$ of x_0 such that $|f_i(x) - f(x)| < \delta$ for all x in $N(\delta,x_0)$ and for all $i \geq I$.

<u>Definition</u> 10.2. A sequence $\{f_i(x), x \in \mathbb{R}^n\}$ is said to have <u>small oscillation</u> at $x = x_0$ if for any $\delta > 0$, there exist $I = I(\delta) > 0$ and a neighborhood $N(\delta,x_0)$ of x_0 such that $|f_i(x) - f_i(x_0)| < \delta$ for all x in $N(\delta,x_0)$ and for all $i \geq I$.

In the following two lemmas we study the convergence of the sequences $\{w_{ij}\}$ to w_∞ defined earlier.

<u>LEMMA</u> 10.3. $\{w_{ij}\}$ converges to w_∞ locally at $x = x_0$ if $\{w_{ij}\}$ has small oscillation at $x = x_0$.

PROOF: Suppose that $\{w_{ij}\}$ has small oscillation at $x = x_0$. Let δ be any positive number. Then there exist $I = I(\delta) > 0$ and n neighborhood $N(\delta,x_0)$ of x_0 such that $|w_{ij}(x) - w_{ij}(x_0)| < \delta$ for all $ij > I$ and $x \in N(\delta,x_0)$. This clearly implies that $|w_\infty(x) - w_\infty(x_0)| < \delta$ for $x \in N(\delta,x_0)$. Choose a rational number x_1 in $N(\delta,x_0)$. Since $w_{ij}(x) \to w_\infty(x)$ pointwise for rational x, there exists $I_1 = I_1(\delta) > 0$ such that $|w_{ij}(x_1) - w_\infty(x_1)| < \delta$ for $ij > I_1$. Thus for $x \in N(\delta,x_0)$ and $ij > \max(I,I_1)$.

$$|w_{ij}(x) - w_\infty(x)| \leq |w_{ij}(x) - w_{ij}(x_0)| + |w_{ij}(x_0) - w_{ij}(x_1)|$$

$$+ |w_{ij}(x_1) - w_\infty(x_1)| + |w_\infty(x_1) - w_\infty(x)|$$

$$\leq \delta + \delta + \delta + \delta = 4\delta.$$

This proves the lemma.   Q.E.D.

COROLLARY 10.4. Suppose that x_0 is a point of continuity for w_∞^+ and w_∞^-. Then $w_{ij} \to w$ locally at x_0. In particular, $w_{ij} \to w$ locally except for countable points.

PROOF: From our hypothesis, given any $\delta > 0$, there exists a closed neighborhood $N(\delta, x_0)$ of x_0 such that $|w_\infty^+(x) - w_\infty^+(x_0)| + |w_\infty^-(x) - w_\infty^-(x_0)| < \delta$. Let $\bar{N}(\delta, x_0)$ be any proper subinterval of $N(\delta, x_0)$ and an open neighborhood of x_0 so that there exist rational numbers $x_1 < x_0$ and $x_2 > x_0$ in $N(\delta, x_0) - \bar{N}(\delta, x_0)$. Since $w_{ij}^\pm(x_1) \to w_\infty^\pm(x_1)$ and $w_{ij}^\pm(x_2) \to w_\infty^\pm(x_2)$, there exists $I = I(\delta) > 0$ such that $|w_{ij}^\pm(x_1) - w_\infty^\pm(x_1)| + |w_{ij}^\pm(x_2) - w_\infty^\pm(x_2)| < \delta$. Thus for any $\bar{x}$ and $\bar{\bar{x}}$ in $\bar{N}(\delta, x_0)$, $x_1 < \bar{x} < \bar{\bar{x}} < x_2$, since $w_{ij}^\pm$ are monotone, we have

$$|w_{ij}^\pm(\bar{x}) - w_{ij}^\pm(\bar{\bar{x}})| \leq |w_{ij}^\pm(x_2 - w_{ij}^\pm(x_1)|$$

$$\leq |w_{ij}^\pm(x_2) - w_\infty^\pm(x_2)| + |w_\infty^\pm(x_2) - w_\infty^\pm(x_1)|$$

$$+ |w_\infty^\pm(x_1) - w_{ij}^\pm(x_1)| \leq 3\delta, \quad ij \geq I.$$

In other words, $\{w_{ij}\}$ has small oscillation at $x = x_0$ and the corollary follows from Lemma 10.3. Q.E.D.

Remark 10.5. When $\{w_{ij}\}$ has small oscillation at $x = x_0$ and yet it has large variation at $x = x_0$, $\{w_{ij}^+\}$ and $\{w_{ij}^-\}$ have large oscillation. In this case $\{w_{ij}\}$ converges to w_∞ locally, but $\{w_{ij}\}$ does not converge to $w_\infty^\pm$ locally. In fact, w_∞^+ and w_∞^- have jumps at $x = x_0$ of the same order. It is also possible that while w_∞ is continuous at $x = x_0$, $w_\infty^\pm$ are not continuous at $x = x_0$ and the convergence of w_{ij} to w_∞ is not local at $x = x_0$. It is noted that although $w_\infty = w_\infty^+ + w_\infty^-$, $T.V.w_\infty$ may not equal $T.V.w^+ + T.V.w_\infty^-$. This corresponds to points of interactions to be discussed in Section 15.

We now turn to the convergnece of approximate solutions $u_r(x,t)$ discussed in the last two sections. Apply the above results to each component $u_r^k(x,t)$, $k = 1,2,\ldots,n$, of $u_r(x,t)$ for each fixed rational t and use diagonal process to show that there exists a sequence $r_i > 0$ $r_i \to 0$ as $i \to \infty$ such that $u_{r_i}^\pm(x,t) \equiv (u_{r_i}^{k\pm},(x,t))$ tend to $u_0^\pm(x,t)$ for each rational t. From Lemma 10.3 and its corollary we obtain local convergence of $u_{r_i}(x,t)$ to $u_0(x,t)$, t rational, as a function of x. We further assume that the measures $dQ(u_{r_i})$ and $dC(u_{r_i})$ tends to $dQ(u_0)$ and $dQ(u_0)$ in weak topology for measures. This is done by applying the diagonal process again since $\{dQ(U_{r_i})\}$ and $\{dC(u_{r_i})\}$ are uniformly bounded measures, Corollary 8.2. We now show that $u_0(x,t)$ can be extended to all $t \geq 0$ and the convergence is local in two independent variables (x,t).

THEOREM 10.6. Suppose that the total variation T.V. of the initial data (1.2) is sufficiently small. Then there exist a sequence $r_i \to 0_+$ and a bounded function $u_0(x,t)$ with the following properties. For each fixed $t \geq 0$, $u_0(x,t)$ is function of bounded variation in x with variation $O(1)$T.V. and the sequence of approximate solutions $\{u_{r_i}(x,t)\}$ tends to $u_0(x,t)$ locally for all (x,t) except on set E. Moreoever, the intersection of E with each horizontal line $t = $ constant is a set consisting of at most countable points. Moreover $\{dQ(u_{r_i})\}$ and $\{dC(u_{r_i})\}$ tend to $dQ(u_0)$ and $dC(u_0)$ in the weak star topology for measures.

PROOF: For each $t_0 \geq 0$, not necessarily rational, by the above arguments, there exists a subsequence $\{r_{ij}\} \equiv \{r_{ij}(t_0)\}$ of $\{r_i\}$ such that $\{u_{r_{ij}}(x,t_0)\}$ tends to a limit function $u_0(x,t_0)$ locally except for a countable x (Corollary 10.4). It remains to show that for x_0 not in this countable set, the original sequence $\{u_{r_i}(x,t)\}$ converges locally to $\{u_0(x,t)\}$ at (x_0,t_0). Choose a state u_* around which the initial data $u(x,0)$ are posed. By a linear transformation, we assume that $u^i = $ constant is parallel to $r_i(u_*)$, $i = 1,2,\ldots,n$. Such a choice of the coordinate (u^i) also satisfies the requirements made in Sections 5 and 6 when we measured the strength of i-waves by the changes in u^i. Another advantage of such a choice is that when an approximate solution $u_r(x,t_0)$ has the property that u_r^{i+} and u_r^{i-}, $i = 1,2,\ldots,n$, have small variation around $x = x_0$, then the total amount of i-waves, $i = 1,2,\ldots,n$, around $x = x_0$ is also small, which is true for our choice of the point (x_0,t_0) provided that $r = r_{ij}$, ij large enough. Thus given any $\delta > 0$, there exists $I = I(\delta) > 0$ and a neighborhood $N(t_0)$ of (x_0,t_0) in $t = t_0$ such that the total amount of waves in $u_{r_{ij}}$, $ij > I$, crossing $N(t_0)$ is less than δ. We further assume that $Q(u_0)$ and $C(u_0)$ have zero pointed measure at (x_0,t_0). Since these measures are finite, this precludes only countable points. Thus by choosing I large enough, there exists a neighborhood N of (x_0,t_0) in the x-t plane such that for $ij > I$, $|Q(u_{r_{ij}})|(N) + |C(u_{r_{ij}})(N)| < \delta$. We may assume that $N \cap \{\text{time} = t_0\} = N(t_0)$. Denote by $N(t)$ the intersection of N with time $= t$. By choosing N properly, we have from the conservation of waves in Lemma 9.2 that the total amount of waves in $u_{r_{ij}}$, $ij > I$, crossing $N(t)$, t close to t_0, is less than 2δ. Pick t_1 rational and close to t_0. We know that not only $\{u_{r_{ij}}(x,t_1)\}$ but also $\{u_{r_i}(x,t_1)\}$ tends to $u_0(x,t_1)$ locally outside countable x. Since the total amount of waves in

$u_{r_{ij}}$, $ij > I$, crossing $N(t_1)$ is less than 2δ, it is clear that the total variation of $u_{r_{ij}}^{\pm}(x,t_1)$ over $N(t_1)$ is also small. This implies that the toal variation of $u_0^{\pm}(x,t_1)$ over $N(t_1)$ is small. By arguments used in the proof of Corollary 10.4, we see that there exists $I_1 > I$, such that the total variation of $u_{r_i}^{\pm}(x,t_1)$, $i > I_1$, over $N(t_1)$ is small. This implies that the total amount of waves in $u_{r_i}^{\pm}$, $i > I$, crossint $N(t_1)$ is less than 3δ. Again by Lemma 9.2, there exists an open set $\bar{N} \subset N$ such that the total amount of waves in u_{r_i}, $i > I_1$, crossing $\bar{N}(t)$ is less than 4δ. By choosing t_1 sufficiently close to t_0, we may assume that $\bar{N}$ is a neighborhood of (x_0,t_0). This shows that $u_{r_i}(x,t)$, $i > I_1$, has an oscillation less than 5δ over N. The proof of the theorem now follows from the arguments used in the proof of Lemma 10.3. Details are omitted.

<u>THEOREM</u> 10.6. The limit function $u_0(x,t)$ in the above theorem is a weak solution of (1.1) and (1.2) provided that the sequence $\{a_k\}$ is equidistributed in $(-1,1)$.

PROOF: We have to show that (1.3) holds for the limit function $u_0(x,t)$. For approximate solution $u_r(x,t)$ we have

$$E(r) \equiv \iint_{t \geq 0} (u_r \frac{\partial \phi}{\partial t} + f(u_r) \frac{\partial \phi}{\partial x}) dxdt + \int_{-\infty}^{\infty} u_r(x,0)\phi(t,0) dx$$

$$= \sum_{k=0}^{\infty} \int_{-\infty}^{\infty} (u_r(x,ks+0) - u_r(x,ks-0)]\phi(x,ks) dt$$

$$\equiv \sum_{k=0}^{\infty} E_k(r).$$

Since the test function ϕ has compact support, the above summation consists of $O(1)\frac{1}{r}$ nonzero terms and each $E_k(r) \equiv O(1)T.V.r$. This immediately yields that $E(r) = O(1)$. To see that $E(r)$ tends to zero as r tends to zero, we use Theorem 9.1 and the equidistributedness of $\{a_k\}$ to show that $\{E_k(r)\}_{k=0}^{\infty}$ cancel each other out if $Q + C \equiv 0$. The nonlinear effects Q and C are minimized by taking advantage of the fact that the subsequence $\{a_k, qN < k < pN\}$ is equidistributed for any fixed $p > q$, as $N \to \infty$. Details are omitted, [31].   Q.E.D.

§11. <u>Expansion waves</u>

In Section 9 we define generalized characteristics for approximate solutions and for each k-characteristic $X_k(u_r)$, we define a Lipschitz continuous curve $\overline{X}_k(u_r)$. Let $\{X_k(u_{r_i})\}$ be a sequence of k-characteristics in a compact set of x-t plane. It follows from Lemma 9.4 that there exists a subsequence $\{r_{ij}\}$ of $\{r_i\}$, such that $\{X_k(u_{r_{ij}})\}$ and $\{\overline{X}_k(u_{r_{ij}})\}$ tend to a Lipschitz continuous curve $X_k \equiv X_k(u_0)$ as r_{ij} tends to zero. The sequence $\{r_{ij}\}$ may, of course, depend on $\{X_k(u_{r_i})\}$. The amount of k-expansion waves (as measured by $[\lambda_k]$ the rate of expansion of these waves) in $u_r(x,t)$ at time t is denoted by $w_k^+(t,u_r)$. For a fixed t = T, we may assume by diagonal process that the sequence of nonnegative measures $\{dw_k^+(T;u_{r_{ij}})\}$ tends to $d\overline{w}_k^+(T) \equiv d\overline{w}_k^+(T;\{u_{r_{ij}}\})$ in the weak* topology for measures as r_{ij} tends to zero. Similarly we have nonpositive measures $d\overline{w}_k^-(T)$ for k-compression waves. For approximate solutions, the amount of k-compression waves is the decreasing variation of λ_k across k-waves. On the other hand, the weak solution $u_0(x,t)$ at a fixed time t = T has bounded total variation in x and thus $du_0(x,T)$ can be decomposed into expansion and compression k-waves $dw_k^\pm(T)$. By the pointwise convergence results proved in the last section, we see that $dw_k^+(T) + dw_k^-(T) = d\overline{w}_k^+(T) + d\overline{w}_k^-(T)$. However, due to the possible cancellation between $d\overline{w}_k^+(T)$ and $d\overline{w}_k^-(T)$, we have $d\overline{w}_k^+(T) \geq dw_k^+(T)$ and $|d\overline{w}_k^-(T)| \geq |dw_k^-(T)|$.

Let X_k^1 and X_k^2 be two k-characteristics obtained by the above limiting process and $\overline{w}_k^+(T_1)$ be the amount of expansion waves in $u_0(x,t)$ at time T_1, and $D_k(t)$ the distance between X_k^1 and X_k^2 at time t. These quantities depend on the particular sequence $\{r_{ij}\}$ used in diagonal process. Nevertheless, the following theorem always holds.

THEOREM 11.1. Suppose that X_k^1 and X_k^2 are defined for $t \in (T_0,T_1)$ and Λ is the region between X_k^1, X_k^2, $t = T_0$ and $t = T_1$. Then

$$w_k^+(T_1) \leq \frac{D(T_1)}{T_0-T_1} \exp[\tilde{w}_k(T_0,T_1) + Q(\Lambda)] + P_{\lambda_i}(\Lambda) + O(1)Q(\Lambda)$$

where $\tilde{w}_k(T_0,t)$, $T_0 \leq t \leq T_1$, denotes the limit sup as $r_{ij} \to 0$ of the total amount of the i-waves $i \neq k$, in approximate solutions $u_{r_{ij}}$ which cross $X_k^1(u_{r_{ij}})$ and $X_k^2(u_{r_{ij}})$ between time T_0 and T_1 or time = T_0 between $X_k^1(u_{r_{ij}})$ and $X_k^2(u_{r_{ij}})$.

PROOF: Let I be the interval between X_k^1 and X_k^2 at time T_1. Given any $\varepsilon > 0$, we may divide I into subintervals $I_1,I_2,\ldots,$ $I_N,I_{N+1},\ldots,I_{N+M}$, so that the total amount of k-expansion waves in $w_k^+(T_1)$ over $I_1,I_2,\ldots,I_N$ is larger than $w_k^+(T_1) - \varepsilon$.

$\bar{\chi}^1_{k,\ell}$ always lies to the left of $\bar{\chi}^2_{k,\ell}$ and Lemma 9.4 holds. The position of these curves are denoted by $(x_1(t),t)$ and $(x_2(t),t)$ and $D_\ell(t)$ $\equiv x_2(t) - x_1(t)$ is the distance between them. Since $\chi^1_{k,\ell}(u_{r_{ij}})$ and $\chi^2_{k,\ell}(u_{r_{ij}})$ are of type I, they are contained in k-rarefaction waves and so

$$\frac{dD_\ell(t)}{dt} = \lambda_k(u^+(t)) - \lambda_k(u^-(t)), \qquad T_0 \le t \le T_1,$$

where $u^+(t)$ and $u^-(t)$ are the values of $u_{r_{ij}}(x,t)$ which $\chi^1_{k,\ell}(u_{r_{ij}})$ and $\chi^2_{k,\ell}(u_{r_{ij}})$ assume at time t. We denote by $w^-_{k,\ell}(t)$ the amount of k-compression waves in $u_{r_{ij}}(x,t)$ between $\chi^1_{k,\ell}(u_{r_{ij}})$ and $\chi^2_{k,\ell}(u_{r_{ij}})$ at time t; and $\sigma_{k,\ell}(t)$ the total amount of i-waves, $i \ne k$, in $u_{r_{ij}}(x,t)$ between $\chi^1_{k,\ell}(u_{r_{ij}})$ and $\chi^2_{k,\ell}(u_{r_{ij}})$ at time t. We have from the above estimate that

(11.3) $$\frac{dD_\ell(t)}{dt} = w^+_{k,\ell}(t) + w^-_{k,\ell}(t) + O(1)\sigma_{k,\ell}(t).$$

Let $\tilde{w}_{k,\ell}(t;u_{r_{ij}})$ be the amount of i-waves, $i \ne k$, crossing $\chi^1_{k,\ell}(u_{r_{ij}})$ and $\chi^2_{k,\ell}(u_{r_{ij}})$ between time t and $T_0 - O(1)D_\ell(T_0)$ and $Q_\ell(t;u_{r_{ij}})$ the amount of interactions in Λ_ℓ between t and $T - O(1)D_\ell(T_0)$. It follows from the hyperbolicity of the system (1.1) that the i-characteristic, $i < k$, (or $i > k$) which meets $\chi^1_{k,\ell}(u_{r_{ij}})$ (or $\chi^2_{k,\ell}(u_{r_{ij}})$ at time t also meets $\chi^2_{k,\ell}(u_{r_{ij}})$ (or $\chi^1_{k,\ell}(u_{r_{ij}})$ before time $s < t$; and, moreover,

$$s = t - O(1)\bar{D}_\ell(t)(1+O(r_{ij}))$$

where $\bar{D}(t)$ is the distance between $\bar{\chi}^1_{k,}(u_{r_{ij}})$ and $\bar{\chi}^2_{k,\ell}(u_{r_{ij}})$ at time t. In the above estimate we have used (ii) of Lemma 9.4 to estimate the position of the i-characteristics. As before, we always choose the sequence $\{a_k\}$ to be equidistributed so that in the estimate (iii) of Lemma 9.4, $A(k) = O(r_{ij})$. Similarly, we also have

$$\bar{D}_\ell(t) = D_\ell(t) + O(r_{ij}), \qquad T_0 \le t \le T_1.$$

The above two estimates yield

(11.4) $s = t - O(1) D_\ell(t) + O(r_{ij})$.

By our choice of s and Lemma 9.2, the total amount of i-waves, $i \neq k$, crossing time $= t$ between $\chi^1_{k,\ell}(u_{r_{ij}})$ and $\chi^2_{k,\ell}(u_{r_{ij}})$ is dominated by the amount of i-waves $i \neq k$, crossing $\chi^1_{k,\ell}(u_{r_{ij}})$ and $\chi^2_{k,\ell}(u_{r_{ij}})$ between time t and time s and the total amount of interactions in the region between $\chi^1_{k,\ell}(u_{r_{ij}})$, $\chi^2_{k,\ell}(u_{r_{ij}})$, time t and s. Thus

$$\sigma_{k,\ell}(t) \leq \tilde{w}_{k,\ell}(t; u_{r_{ij}}) - \tilde{w}_{k,\ell}(s; u_{r_{ij}}) + O(1)[Q_\ell(t; u_{r_{ij}}) - Q_\ell(s; u_{r_{ij}})].$$

We have from (11.3), (11.4) and the above estimate that

$$\frac{dD_\ell(t)}{dt} = w^+_{k,\ell}(t) + w^-_{k,\ell}(t) + O(1) \int_{t-O(1)D_\ell(t) + O(r_{ij})}^t d[\tilde{w}_{k,\ell}(\tau; u_{r_{ij}}) + Q_\ell(\tau; u_{r_{ij}})].$$

Integrating the above estimate and changing the order of the resulting double integration, we obtain

$$D_\ell(t) = D_\ell(T_0) + \int_{T_0}^t [w^+_{k,\ell}(\tau) + w^-_{k,\ell}(\tau)]d\tau + O(r_{ij})$$

$$+ O(1) \int_{T_0 - O(1)D_\ell(T_0)}^t D_\ell(\tau) d(\tilde{w}_{k,\ell}(\tau; u_{r_{ij}}) + Q_\ell(\tau; u_{r_{ij}})),$$

$$T_0 \leq t \leq T_1,$$

and so,

$$D_\ell(t) = D_\ell(T_0)[1 + O(1)(\tilde{w}_{k,\ell}(T_0; u_{r_{ij}}) + Q_\ell(T_0; u_{r_{ij}}))]$$

$$+ \int_{T_0}^t [w^+_{k,\ell}(\tau) + w^-_{k,\ell}(\tau)]d\tau + O(r_{ij})$$

$$+ O(1) \int_{T_0}^t D_\ell(\tau) d(\tilde{w}_{k,\ell}(\tau; u_{r_{ij}}) + Q_\ell(\tau; u_{r_{ij}})), \quad T_0 \leq t \leq T_1.$$

The above estimate is an integral inequality for $D_\ell(\tau)$ which can be solved to yield:

(11.5) $D_\ell(T_1) \geq D_\ell(T_0)[1 + O(1)h_\ell(T_0)] + O(r_{ij})$

$$+ \int_{T_0}^{T_1} [w_{k,\ell}^+(\tau) + w_{k,\ell}^-(\tau)][2 - \exp O(1)(h_\ell(T_1) - h_\ell(\tau))]d\tau,$$

$$h_\ell(\tau) \equiv \tilde{w}_{k,\ell}(\tau; u_{r_{ij}}) + Q_\ell(\tau; u_{r_{ij}}).$$

Since we are dealing with weak waves, h_ℓ is small (c.f. (i), Lemma 9.4) and so $1 + O(1)h_\ell(T_0)$ and $2 - \exp O(1)(h_\ell(T_1) - h_\ell(\tau))$ are positive. Thus the first term on the right hand side of (11.5) can be depleted. Note that since $\chi_{k,\ell}^1(u_{r_{ij}})$ and $\chi_{k,\ell}^2(u_{r_{ij}})$ are both of type I, it follows from the results on wave interactions in Section 7, in particular Theorem 7.4, that

$$w_{k,\ell}^+(\tau) + w_{k,\ell}^-(\tau) = w_{k,\ell}^+(T_1) + w_{k,\ell}^-(T_1) + O(1)Q(\Lambda_\ell; u_{r_{ij}}).$$

We have from (11.5) and the above estimate that

(11.6) $w_{k,\ell}^+(T_1) + w_{k,\ell}^-(T_1)$

$$\leq \frac{D_\ell(T_1)}{T_1 - T_0} \exp O(1)[h_\ell(T_1) - h_\ell(T_0)] + O(r_{ij}) + O(1)Q(\Lambda_\ell, u_{r_{ij}}).$$

We note that as $r_{ij} \to 0$, $h_\ell(T_1) - h_\ell(T_0)$, $\ell = 1, 2, \ldots, N$, is dominated by $\tilde{w}_k(T_0, T_1) + Q(\Lambda)$ and

$$\lim_{r_{ij} \to 0} \sum_{\ell=1}^N Q(\Lambda_\ell; u_{r_{ij}}) \leq Q(\Lambda),$$

$$\lim_{r_{ij} \to 0} \sum_{\ell=1}^N D_\ell(T_1) = \lim_{r_{ij} \to 0} \sum_{\ell=1}^N \bar{D}_\ell(T_1) \leq D(T_1)$$

$$\lim_{r_{ij} \to 0} \sum_{\ell=0}^N P_{\lambda_i}(\Lambda_\ell, u_{r_{ij}}) \leq P_{\lambda_i}(\Lambda).$$

Thus it follows from (11.1), (11.2) and (11.6) that

$$w_k^+(T_1) \leq \frac{D(T_1)}{T_1 - T_0} \exp O(1)[\tilde{w}_k(T_0, T_1) + Q(\Lambda)] + P_{\lambda_i}(\Lambda) + O(1)Q(\Lambda) + 4\varepsilon.$$

This proves the theorem since $\varepsilon > 0$ is arbitrary.   Q.E.D.

§12. Continuity points

Our purpose in this section is to clarify more precisely the point of continuity for the weak solution and improve the convergence result in Theorem 10.6. As before we assume that a sequence of approximate solutions $\{u_{r_i}\}$ tends to a weak solution u_0 as $r_i \to 0$, and $dQ(u_{r_i})$ and $dC\,u_{r_i})$ tend to $dQ(u_0)$ and $dC(u_0)$ respectively in the weak star topology for measures.

THEOREM 12.1. Suppose $u_0(x,t_0)$ is continuous with respect to x at (x_0,t_0) and dQ and dC have zero pointed measure at (x_0,t_0). Then $u(x,t)$ is continuous in (x,t) at (x_0,t_0) and $\{u_{r_i}(x,t)\}$ tends to $\{u_0(x,t)\}$ locally at (x_0,t_0) in the sense of Definition 10.1.

PROOF: By Lemma 10.3 and the convergence result in Theorem 10.6, we need only to show that $u_{r_i}(x,t)$ has small oscillation at (x_0,t_0). Our first step is to show that $\{u_{r_i}(x,t_0)\}$ has small oscillation in x at $x = x_0$. Since $dQ + dC$ has zero point measure at (x_0,t_0) for any given $\varepsilon > 0$ there exists a neighborhood N of (x_0,t_0) and $I_0 > 0$ such that

$$(12.1) \qquad\qquad Q(u_{r_i}) + C(u_{r_i})|_N \leq \varepsilon^3, \qquad i > I_0.$$

We now partition the waves in u_{r_i}, $i > I$, for the region $\Lambda = N$ according to Theorem 9.1. Due to the strict hyperbolicity of the system (1.1), there exists a neighborhood $N_0 \subset N$ of (x_0,t_0) with the property that any j-characteristic and k-characteristic, $j \neq k$, intersecting N_0 must meet in N. Consequently, any j-waves and k-waves, $j \neq k$, of type I and II in N_0 must meet in N provided I_0 is sufficiently large (Theorem 9.1) and therefore contribute to the interaction measure $dQ(u_{r_i})$. As a result we have from Theorem 9.1 and the estimate (12.1) that in u_{r_i} there exists k, $1 \leq k \leq n$, such that all j-waves, $j \neq k$, are small. More precisely, we have

$$(12.2) \qquad\qquad \sum_{\substack{j=1 \\ j\neq k}}^{n} x_j(u_{r_i})\Big|_{N_0}(t) \leq \varepsilon, \qquad i > I_0$$

where $N_0(t)$ is the intersection of N_0 with time $= t$, and $x_j(u_{r_i})$ is the amount of j-waves in u_{r_i}. Since $u_0(x,t_0)$ is continuous at $x = x_0$ and has bounded variation, we may choose N sufficiently small that the total variation of $u_0(x,t_0)$ over $N(t_0)$ is less than ε. Thus it follows from the pointwise convergence result in Theorem 10.6 that for I_0 sufficiently large

$$(12.5) \qquad |u_{r_i}(x_-,t_0) - u_{r_i}(x_+,t_0)| \;<\; \varepsilon, \qquad i > I_0$$

where (x_-,t_0) and (x_+,t_0) are the end points of the interval $N_0(t_0)$.

We now prove by contradiction that the sequence $\{u_{r_i}(x,t_0)\}$ has oscillation $0(\varepsilon)$ over the interval $N_0(t_0)$. For this we apply Lemma 6.4. Although the lemma deals with only one family of elementary waves, easy continuity arguments show that the lemma still holds provided that the total amount of elementary waves which do not belong to a particular family is small. Thus from (12.2) and (12.5) we need only to show that when I_0 is sufficiently large

$$w_k^+(u_{r_i}) + w_k^-(u_{r_i})\big|_{L_0} \;>\; -2\sqrt{\varepsilon}, \qquad i > I_0,$$

for any subinterval L_0 of $N_0(t_0)$. We claim that the above follows from

$$(12.6) \qquad w_k^+(u_{r_i}) + w_k^-(u_{r_i})\big|_{L} \;<\; \sqrt{\varepsilon}, \qquad i > I_0$$

for any subinterval L of $N_0(t_0)$. In fact, if $w_k^+(u_{r_i}) + w_k^-(u_{r_i})\big|_{L_0} < -2\sqrt{\varepsilon}$ for some L_0, then (12.2) and 12.5) imply that $w_k^+(u_{r_i}) + w_k^-(u_{r_i})\big|_{L} \geq -2\varepsilon + 2\sqrt{\varepsilon} > \sqrt{\varepsilon}$ for $L = L_1$ or $L = L_2$, which contradicts (12.6). Here L_1 and L_2 are the intervals in $N_0(t_0) - L_0$. Suppose that (12.6) does not hold. Then there exists a sequence $\{r_{ij}\} \subset \{r_i\}$, $r_{ij} \to 0$, and a sequence of $\{L_{ij}\}$ of subintervals of $N_0(t_0)$ such that

$$(12.7) \qquad w_k^+(u_{r_{ij}}) + w_k^-(u_{r_{ij}})\big|_{L_{ij}} \;>\; \sqrt{\varepsilon}, \qquad \text{for all } r_{ij}.$$

Let $\chi_k^1(u_{r_{ij}})$ and $\chi_k^2(u_{r_{ij}})$ be the first and last, respectively, k-characteristics of type I which pass through L_{ij}. We have from Theorem 9.1, (12.7) and (12.1) that

$$(12.8) \qquad w_k^+(u_{r_{ij}}) + w_k^-(u_{r_{ij}})\big|_{\tilde{L}_{ij}} \;>\; 2\varepsilon$$

where $\tilde{L}_{ij}(\subset L_{ij})$ is the interval between $\chi_k^1(u_{r_{ij}})$ and $\chi_k^2(u_{r_{ij}})$. Let

t_1 ($< t_0$) be the time before which $\chi_k^1(u_{r_{ij}})$ and $\chi_k^2(u_{r_{ij}})$ do not leave
N. We have from (12.1), Corollary 8.2, Theorem 11.1 and its proof that

$$(12.9) \quad w_k^+(u_{r_{ij}}) + w_k^-(u_{r_{ij}})\big|_{\tilde{L}_{ij}} \leq \frac{m}{t_0-t_1}\exp(0(1)\varepsilon^3) + 0(1)\varepsilon^3 + 0(r_{ij})$$

where m is the length of $\tilde{L}_{ij}$. By choosing the set N_0 to be sufficiently small we can easily make

$$(12.9) \qquad\qquad \frac{m}{t_0-t_1} < \varepsilon$$

and so for small r_{ij}, $0(r_{ij}) < \varepsilon$, we have from (12.9) that (12.8) does
not hold. In other words, (12.6) holds and $\{u_{r_i}(x,t_0)\}$ has oscillation
$0(\varepsilon)$ over $N_0(t_0)$.

Finally, we show that $\{u_{r_i}\}$ has small oscillation in (x,t) at
(x_0,t_0). The above arguments may be applied to show that for N_0 sufficiently small, $\{u_{r_i}(x,\tau)\}$ has oscillation $0(\varepsilon)$ in x over $N_0(\tau)$ for
any τ. It remains to show that $\{u_{r_i}(x,t)\}$ has small oscillation along
a time-like curve. Choose any curve $x = \chi(u_{r_i})$ in N_0 which does not
propagate with any characteristic speed. Thus an elementary wave crosses
χ does so transversally. We have from Theorem 9.1 that the wave pattern
along χ is close to those along $N(t_0)$. Thus similar estimates as (12.2),
(12.6) hold along χ and so we may apply Lemma 6.4 to conclude that
$\{u_{r_{ij}}\}$ has oscillation $0(\varepsilon)$ along χ. This completes the proof of the
theorem.   Q.E.D.

The above proof yields easily the following theorem.

THEOREM 12.2. Suppose that (x_0,t_0) is a continuity point described in
Theorem 12.1. Then in a small neighborhood of (x_0,t_0) the weak solution
$u_0(x,t)$ has small variation in x for (x,t) near (x_0,t_0) and also has
small variation along any time-like curve near (x_0,t_0).

§13. Curves of discontinuity

The purpose of this section is to study the curves of discontinuity in
the weak solution. We remind the reader that there are two kinds of k-
discontinuities: An admissible k-discontinuity (u_-,u_+) simple if for
any u on $S_k(u_-)$ strictly between u_- and u_+, $\sigma(u_-,u_+) < \sigma(u_-,u)$.
When (u_-,u_+) is nonsimple and $\sigma(u_-,u_+) = \sigma(u_-,u_1) = \sigma(u_-,u_2) = \cdots = $
$\sigma(u_-,u_\ell)$ for a monotone sequence of states $u_1,u_2,\ldots,u_\ell$ on $S_k(u_-)$

strictly between u_- and u_+, we say that (u_-, u_+) is a <u>composite</u> of $(u_-, u_1), (u_1, u_2), \ldots, (u_\ell, u_+)$. It is clear that $(u_-, u_1), (u_1, u_2), \ldots, (u_\ell, u_+)$ are admissible k-discontinuities; and (u_-, u_1), (u_ℓ, u_+), and (u_i, u_{i+1}), $i = 1, 2, \ldots, \ell-1$, are, respectively, right-sided, left-sided, and two-sided contact continuities.

<u>THEOREM</u> 13.1. Suppose that $u_0(x, t_0)$ is discontinuous at $x = x_0$ and $dQ + dC$ has zero point measure at (x_0, t_0). Then $(u_-, u_+) \equiv (u(x-0, t_0),$ $u(x+0, t_0))$ is an admissible k-discontinuity for some $k \in \{1, 2, \ldots, n\}$. When (u_-, u_+) is simple, there exists a Lipschitz continuous curve Γ in a neighborhood of (x_0, t_0) with the following properties. At (x_0, t_0), Γ propagates with shock speed $\sigma(u_-, u_+)$. $u_0(x, t)$ has small variation in x in $N - \Gamma$, N a small neighborhood of (x_0, t_0). Moreover, for r_i sufficiently small, there exists a k-discontinuity $\Gamma(u_{r_i})$ in u_{r_i} such that $\{\Gamma(u_{r_i})\}$ tends to Γ as r_i tends to zero and $\{u_{r_i}(x, t), (x, t)$ to the left (or right) of $\Gamma(u_{r_i})\}$ has small oscillation at (x_0, t_0). When (u_-, u_+) is a composite of $(u_-, u_1), (u_1, u_2), \ldots, (u_\ell, u_+)$, there exist Lipschitz continuous curves $\Gamma_1, \ldots, \Gamma_\ell$ in a neighborhood of (x_0, t_0) such that these curves propagate with speed $\sigma(u_-, u_+)$ at (x_0, t_0) and outside these curves $u_0(x, t)$ has small variation in small neighborhood of (x_0, t_0). The curve Γ_j and Γ_{j+1}, $j = 1, \ldots, \ell-1$, may be identical; when there are not identical, Γ_{j+1} lies to the right of Γ_j and between Γ_j and Γ_{j+1}, $u_0(x, t)$ tends to u_j as (x, t) tends to (x_0, t_0). Moreover, for r_i sufficiently small there exists i-discontinuities, $\Gamma_1(u_{r_i}), \ldots,$ $\Gamma_\ell(u_{r_i})$, some of them may be identical, such that $\Gamma_j(u)$, $j = 1, \ldots, \ell$ tend to Γ_j and $\{u_{r_i}(x, t), (x, t)$ between $\Gamma_j(u_{r_i})$ and $\Gamma_{j+1}(u_{r_i})\}$, $j = 1, 2, \ldots, \ell-1$, $\{u_{r_i}(x, t), (x, t)$ to the left of $\Gamma_1(u_{r_i})\}$ and $\{u_{r_i}(x, t), (x, t)$ to the right of $\Gamma_\ell(u_{r_i})\}$ have small oscillation in a small neighborhood of (x_0, t_0).

PROOF: Given $\varepsilon > 0$, we may choose an open neighborhood N of (x_0, t_0) and $I_0 > 0$ such that

(13.1)
$$|Q(u_{r_i}) + C(u_{r_i})|\big|_N < \varepsilon, \qquad i > I_0.$$

Let N_0 be a subset of N and an open neighborhood of N. For small N_0, we have from (13.1) (c.f. (12.2)) that for some $k \in \{1, 2, \ldots, n\}$,

(13.2)
$$\sum_{\substack{j=1 \\ j \neq k}}^{n} X_j(u_{r_i})\big|_{N_0}(t) \leq \varepsilon, \qquad i > I_0.$$

By choosing N small enough, we may assume that $u_0(x,t_0)$ has small total variation in $N - N_0$. Thus it follows from (13.1) and the arguments in the proof of Theorem 12.1 that for I_0 sufficiently large

$$(13.3) \qquad \text{Osc.}_x \, u_{r_i}(x,t_0)\big|_{N-N_0} < \varepsilon, \qquad i > I_0.$$

We draw k-characteristics backward in time through points in $N_0(t)$ and apply Theorem 11.1 to obtain that (c.f. (12.6))

$$(13.4) \qquad w_k^+(u_{r_i}) + w_k^-(u_{r_i})\big|_L < \sqrt{\varepsilon}, \quad i > I_0$$

for any subinterval L of $N_0(t_0)$. Let $\chi_k^1(u_{r_i})$ and $\chi_k^2(u_{r_i})$ be the first and last (from the left) k-characteristics of type I or II through $N_1(t_0)$. Here $N_1(\subset N_0)$ is a small neighborhood of (x_0,t_0). From (13.3), (13.1) and Lemma 9.2 on conservation of waves, we see that

$$(13.5) \qquad \text{Osc.}_x \, u_{r_i}(x,t)\big|_{N-\Lambda} < \varepsilon, \qquad i > I_0$$

where Λ is the region between $\chi_k^1(u_{r_i})$ and $\chi_k^2(u_{r_i})$. The speed of $\bar{\chi}_k^1(u_{r_i})$ changes due to (i) interactions of waves of different families and cancellations along $\chi_k^1(u_{r_i})$, (ii) i-waves, $i \neq k$, crossing $\chi_k^1(u_{r_i})$, and (iii) combining of k-discontinuities along $\chi_k^1(u_{r_i})$, Theorem 7.4. The effect of (i) is ε by (13.1) and that of (ii) is also ε by (13.2). Because of (13.5), except for an error of the order ε, the effect of (iii) is due to the combining of the k-discontinuity on $\chi_k^1(u_{r_i})$ with k-discontinuities from the right. It follows from (v) of Theorem 7.4 that the last effect always decreases the speed of $\bar{\chi}_k^1(u_{r_i})$. Analogous arguments also hold for $\bar{\chi}_k^2(u_{r_i})$ and we conclude that the speed $\sigma^1(t)$ and $\sigma^2(t)$ of $\bar{\chi}_k^1(u_{r_i})$ and $\bar{\chi}_k^2(u_{r_i})$, respectively, at time t satisfy

$$(13.6)_1 \qquad\qquad \sigma^1(t_2) \leq \sigma^1(t_1) - O(1)\varepsilon$$

$$(13.6)_2 \qquad\qquad \sigma^2(t_2) \geq \sigma^2(t_1) + O(1)\varepsilon, \quad t_2 \geq t_1.$$

Here we have assumed that $\chi_k^1(u_{r_i})$ and $\chi_k^2(u_{r_i})$ are contained in N_0 at time t_1 and t_2. By the stability of wave patterns, Theorem 9.1, and (13.1) we see from (13.4) that

(13.7) $\qquad\qquad\qquad w_k^+(u_{r_i}) + w_k^-(u_{r_i})\big|_L < 2\sqrt{\varepsilon}$

for any subinterval L of $N_0(t)$, t any. The above estimate and (13.2) imply

(13.8) $\qquad\qquad\qquad \sigma_2(t) - \sigma_1(t) \le O(1)\sqrt{\varepsilon}.$

In view of (13.6) and (13.7), the relative position of $\overline{\chi}_k^1(u_{r_i})$ and $\overline{\chi}_k^2(u_{r_i})$, except for an error of the order $O(1)\sqrt{\varepsilon}$, may be depicted in Figure 13.1

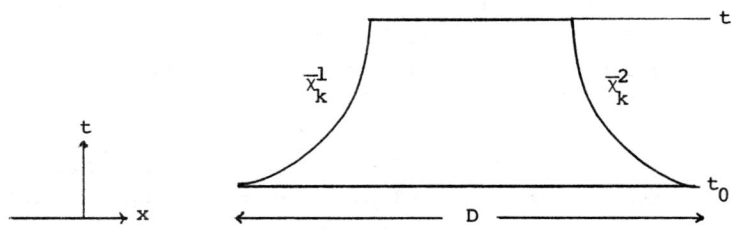

Figure 13.1

It follows from (13.6), (13.8) and Lemma 9.4 that

$$-O(1)\sqrt{\varepsilon} \le \sigma_1(t) - \sigma_2(t) \le \frac{D}{t-t_0} + O(1)\sqrt{\varepsilon} + 0(r_i), \qquad i \ge I_0$$

for all t ($\ge t_0$) for which χ_k^1 and χ_k^2 are in N_0. Here D is the distance between $\chi_k^1(u_{r_i})$ and $\chi_k^2(u_{r_i})$ at time t_0. By choosing N_1 small as compared to N_0 we may assume $t - t_0$ to be large as compared to D and so for I_0 sufficiently large we have $0(r_i)$ small and the above estimate yields:

(13.9) $\qquad\qquad\qquad |\sigma_1(t_1) - \sigma_2(t_1)| = O(1)\sqrt{\varepsilon}$

for some $t_1 \ge t_0$ so that $\chi_k^1(u_{r_i})$ and $\chi_k^2(u_{r_i})$, $i \ge I_0$, lie in N_1 between t_0 and t_1. We claim that, as a consequence of (13.7) and (13.9), the amount of compression k-waves between $\chi_k^1(u_{r_i})$ and $\chi_k^2(u_{r_i})$ at time t_1 is small. More precisely,

(13.10) $w_k^+(u_{r_i}) + w_k^-(u_{r_i})\big|_{L_1} = O(1)\sqrt{\epsilon}$

for any subinterval of $N_0(t_1)$ which lies between $\chi_k^1(u_{r_i})$ and $\chi_k^2(u_{r_i})$.
Indeed, (13.2) and (13.9) imply that the total amount of expansion and
compression k-waves between $\chi_k^1(u_{r_i})$ and $\chi_k^2(u_{r_i})$ at time t_1 is
$O(1)\sqrt{\epsilon}$. Thus, if (13.10) fails then (13.7) would also fail for L being
one of the complement of L_1 with respect to the interval between $\chi_k^1(u_{r_i})$
and $\chi_k^2(u_{r_i})$ at time t_0. This proves (13.10). Thus by continuity arguments using (13.2), we may apply Lemma 6.4 to conclude that, except for an
error of $O(1)\sqrt{\epsilon}$, $u_1(t_1)$ and $u_2(t_1)$ are related by k-waves which form
a solution of a Riemann problem. Here $u_1(t_1)$ (or $u_2(t_1)$ is the limit
of u_{r_i} from the left (right) of $\chi_k^1(u_{r_i})$ ($\chi_k^2(u_{r_i})$ at time t_1. From
(13.5) and the pointwise convergence results proved in Section 12, the
above statement also holds when $u_1(t_1)$ and $u_2(t_2)$ are replaced by u_-
and u_+, respectively. Thus we have shown, in particular, that u_- and
u_+ are related by a noninteracting wave pattern of k-waves, i.e., $u_+ \in$
$T_k(u_-)$. On the other hand, (13.7) means that this wave pattern contains
no expansion waves. Consequently, u_- and u_+ are related by a single
k-discontinuity. This proves the first statement of the theorem.

We now prove the rest of the theorem when (u_-,u_+) is simple. The
above application of Lemma 6.4 and continuity arguments show that by taking
ϵ sufficiently small, when r_i is small, there exists a single k-discontinuity $\Gamma(u_{r_i})$ in u_{r_i} at time t_1 such that the oscillation of u_{r_i}
near (x_0,t_1) outside $\Gamma(u_{r_i})$ is small. Next we show that this holds not
only at time t_1, but for all time t near t_0. For this we need only
show, by arguments using Lemma 6.4 and (13.2) as above, that waves at time
t, $|t-t_0|$ small, and near (x_0,t_0) is weakly interacting so that the
hypothesis of Lemma 6.4 is satisfied. In the present situation we need to
prove that after waves in N are partitioned with respect to ϵ, the following holds:

Claim: For any $t \leq t_1$ and $(x,t) \in N_0$ there do not exist k-waves
on the left in $u_{r_i}(x,t)$ with values around u_- and total strength no
less than $\epsilon^{1/4}$, and k-waves on the right in $u_{r_i}(x,t)$ with values
around u_+ and strength no less than $\epsilon^{1/4}$ such that these two groups of
k-waves make an angle no less than $\epsilon^{1/4}$.

Indeed, due to (13.7), $u_0\big|_{N_0}$ consists mainly of compression waves,
and so the above claim would imply that analogous assumption as in Lemma
6.4 holds. The claim is proved by contradiction in two steps. First, we
see that when the claim fails, then the amount of potential wave interactions

as defined at the beginning of Section 8 between $\chi_k^1(u_{r_i})$ and $\chi_k^2(u_{r_i})$ at

time $t \leq t_0$ is no less than $\varepsilon^{3/4}$. Because between $\chi_k^1(u_{r_i})$ and

$\chi_k^2(u_{r_i})$ at time t_1, u_{r_i} consists of a dominating k-discontinuity as

we have just shown, the potential amount of wave interactions in N_0 at

time t_1 is less than ε. Thus if we can show that the difference between

the potential amounts of wave interactions in N_0 at time t and time t_1

equals (mod $O(1)\varepsilon$) the amount of actual wave interactions in N_0 between

t and t_1, then $Q(N_0)$ would be larger than $\varepsilon^{3/4} - O(1)\varepsilon > \varepsilon$ provided

that the claim fails. This would contradict (13.1) and proves the claim.

We know from (8.1) and (i) of Corollary 8.2 that the difference between the

potential amounts of interactions at two times always dominates the amount

of actual interactions between these times. However, the remark at the

beginning of Section 8 indicates that the two amounts are actually the same

(mod $Q(N_0)$) provided that the waves in N_0 consist mainly of compression

waves, which is the case here. This proves the claim. We have thus shown

that there exists a single k-discontinuity $\Gamma(u_{r_i})$ in u_{r_i} in N_0 with

the property that the oscillation of u_{r_i} on either side of $\Gamma(u_{r_i})$ near

(x_0,t_0) is small. It remains to show that $\{\Gamma(u_{r_i})\}$ tends to a Lipschitz

continuous curve Γ which is an admissible discontinuity curve in u_0 and

u_0 has small variation outside Γ near (x_0,t_0). By compactness argu-

ments, this is so for a subsequence $\{\Gamma(u_{r_{ij}})\}$ of $\{\Gamma(u_{r_i})\}$ (c.f. Lemma

9.4). Suppose that it is not true for the sequence $\{\Gamma(u_{r_i})\}$ itself.

Then there exists another subsequence $\{\Gamma(\tilde{u}_{r_{ij}})\}$ which converges to $\tilde{\Gamma}$

with similar properties and $\Gamma \neq \tilde{\Gamma}$ and thus $u_0(x,t)$ has small variation

on either side of $\tilde{\Gamma}$. This is a contradiction because in a small neighbor-

hood of (x_0,t_0) the jump across Γ in u_0 dominates the variation of

u_0. This proves the theorem when (u_-,u_+) is simple.

When (u_-,u_+) is a composite of $(u_-,u_1),\ldots,(u_\ell,u_+)$, we may also

use the above arguments based on Lemma 6.4 to show that the amount of k-

expansion waves in N_0 is small and at time t_1 the wave pattern is

weakly interacting. Then it follows from the reasoning in the remark at

the beginning of Section 8 that the wave pattern in N_0 is weakly inter-

acting. Thus by Lemma 6.4 and the continuity argument, the wave pattern of

u_{r_i} in N_0 is close to a noninteracting wave pattern with no expansion

waves. By partitioning the waves with respect to ε, ε sufficiently

small, we may, according to Lemma 6.4, define characteristic curves

$\Gamma_\ell(u_{r_i}),\ldots,\Gamma_\ell(u_{r_i})$ of type II corresponding to each of i-discontinuities

$(u_-,u_1),\ldots,(u_\ell,u_+)$. Some of these curves may be identical in their posi-

tion. When they are not, the oscillation between them is small. The con-

vergence of these curves to Lipschitz continuous curves as $\{u_{r_i}\}$ tends

to u_0 is verified as above. The proof of the theorem is complete. Q.E.D.

j odd, the solution $u_0(x,t)$ tends pointwise to the rarefaction wave
$(u_{i,j+1}^-, u_{i,j+2})$ as $(x,t) \in \Omega_i^+$ tends to (x_0,t_0). The incoming waves in
Ω_i^- consist of compressive i-waves, that is, as t tends to t_0 from
below, the amount of i-expansion waves in Ω_i^- tends to zero. Finally,
when there exists i, $1 \leq i \leq n$, such that the limiting amount of k-
waves, $k \neq i$, in Ω_k^- is zero at (x_0,t_0), then the i-waves in Ω_i^-
is a genuine compression wave at (x_0,t_0) in the sense that there does not
exist a single i-discontinuity Γ in Ω_i^- with the property that close
to (x_0,t_0) the total amount of i-waves outside Γ is small.

PROOF: Given $\varepsilon > 0$, we may choose small neighborhoods N_0, N_1, N_2
of (x_0,t_0), $N_0 \subset N_1 \subset N_2$, and $I_0 > 0$ such that

(14.1) $$Q(u_{r_i}) + C(u_{r_i})\big|_{N_2-N_0} \leq \varepsilon^3, \qquad i \geq I_0.$$

For convenience the boundary of these neighborhoods are intervals with
slope $dx/dt = r/s$. By choosing N_1 small as compared to N_2, and I_0
sufficiently large, it follows from (14.1) and the arguments in the proof
of Theorem 12.1 that

(14.2) $$\text{Osc.}\{u_{r_i}(x,t_0) : (x,t_0) \in N_1 - N_0\} \leq \varepsilon.$$

We partition the elementary waves in N_2 with the given ε, then construct the first (last) k-characteristic $\chi_k^1(u_{r_i})$ $(\chi_k^2(u_{r_i})$ of type I or II
from the left which intersects N_0 at time t_0. We define the region
$\Omega_k^+(u_{r_i})$ (or $\Omega_k^-(u_{r_i})$ to be the region in $N_1 \cap \{(x,t) : t \geq t_0\}$ (or
$N_1 \cap \{(x,t) : t \leq t_0\}$ which is between $\chi_k^1(u_{r_i})$ and $\chi_k^2(u_{r_i})$ but not between $\chi_j^1(u_{r_i})$ and $\chi_j^2(u_{r_i})$ for $j \neq k$. The regions $G_i^\pm(u_{r_i})$ are defined
as the components in $N_2 - \bigcup_{k=1}^{n} \Omega_k^\pm(u_{r_i}) - N_0$, Figure 14.2.

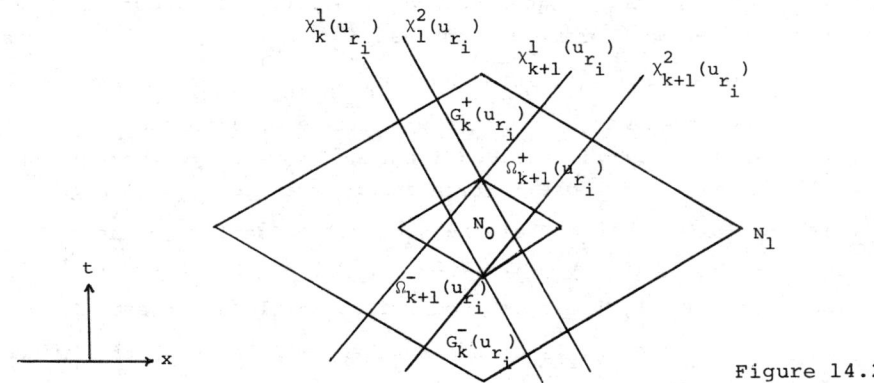

Figure 14.2

(In Figure 14.2, for simplicity, we have depicted characteristics as continuous curves.)

We now study the approximate solution u_{r_i}, $i \geq I_0$, in the region $G_k^-(u_{r_i})$. We claim that $u_{r_i} | G_k^-(u_{r_i})$ has oscillation less than $0(\varepsilon)$. This is the case if the total amount of elementary waves in $G_k^-(u_{r_i})$ is less than ε. Suppose that there exists $\ell \neq k$ such that the total amount of ℓ-waves in $G_k^-(u_{r_i})$ at certain time t_* is larger than ε. For definiteness we assume that $\ell < k$. Let $\chi_\ell^*(u_{r_i})$ be the first characteristics in N_1 of type I or II from the left. We claim that as a consequence of (14.1) the total amount of i-waves, $i \neq \ell$, in N_1, between $\chi_k^*(u_{r_i})$ and $\chi_k^1(u_{r_i})$ is less than ε. This is so because any such i-waves of type I and II would intersect ℓ-waves in $G_k^-(u_{r_i})$ of type I and II and contribute to $Q(u_{r_i}) |_{N-N_0}$. Therefore, we may apply the arguments in the proof of Theorem 12.1 to the region between $\chi_\ell^*(u_{r_i})$ and $\chi_k^1(u_{r_i})$ using (14.1) and (14.2) to conclude that the oscillation of u_{r_i} in this region, and, in particular, in $G_k^-(u_{r_i})$, is small. For such arguments to be applicable, we assume that N_1 is small as compared to N_2.

The above arguments also yield that $u_{r_i} | G_k^+(u_{r_i})$ has small oscillation.

Next we claim that the total amount of k-expansion waves in $\Omega_k^-(u_{r_i})$ is small. Suppose first that the total amount of ℓ-waves, $\ell \neq k$, in $\Omega_k^-(u_{r_i})$ is small. Then the claim is a consequence of Theorem 11.1 provided that N_0 is small as compared to N_1 (c.f. proof of Theorem 12.1). When there exists $\ell \neq k$, such that the total amount of ℓ-waves in $\Omega_k^-(u_{r_i})$ is larger than ε, then we have from the arguments in the previous paragraph that the total amount of k-waves in $\Omega_k^-(u_{r_i})$ is less than ε and $u |_{\Omega_i(u_{r_i})}$ has small oscillation and our claim still holds.

Finally, we claim that $u_{r_i} |_{\Omega_k^+(u_{r_i})}$ is dominated by the elementary waves in (u_{k-1}, u_k). By the uniqueness and stability (c.f. Section 6) of the solution or the Riemann problem, and that $Osc.u |_{G_i^+} = 0(\varepsilon)$, which has just been shown, we need only to show that $u_{r_i} |_{\Omega_k^+(u_{r_i})}$ has a wave pattern close to that of centered waves. When there exists an amount of ℓ-waves, $\ell \neq k$ larger than ε, then it follows from the previous two paragraphs that $u_{r_i} |_{\Omega_k^+(u_{r_i})}$ has $0(\varepsilon)$ oscillation and we are done. We now generalized the arguments in the proof of Theorem 13.1 to deal with the case where k-waves dominate in $\Omega_k^+(u_{r_i})$. Let $\chi_k^-(u_{r_i})$ (or $\chi_k^+(u_{r_i})$) be the first (or last) k-characteristic of type I from the left between $\chi_k^1(u_{r_i})$ and $\chi_k^2(u_{r_i})$. Let t_1 (or t_2) be the smallest (or largest) time which is larger than t_0 and between which χ_k^1, χ_k^2, χ_k^- and χ_k^+ lie in $N_1 - N_0$. The distance between χ_k^1 and χ_k^2 (or between χ_k^- and χ_k^+) at time t is denoted by $D_k(t)$ (or $\overline{D}_k(t)$). By choosing N_0 to be small as compared to N_1, we may assume that

(14.3) $D_k(t_1) \leq \varepsilon(t_2-t_1)$.

By our choice of $\chi_k^-(u_{r_i})$ and $\chi_k^+(u_{r_i})$ (c.f. Section 9 on partition of waves), it follows from (14.1) that between χ_k^1 and χ_k^- and also between χ_k^+ and χ_k^2, u_{r_i} consists mostly of k-compression waves. Thus in these two regions the situation is similar to that of Section 13 and we may use the arguments in the proof of Theorem 13.1 along with (14.1)~(14.3) to conclude that except for a single left-sided (or right-sided) contact dis-continuity, the k-waves between χ_k^1 and χ_k^- (or between χ_k^+ and χ_k^2) at time t_2 consists mainly of two-sided contact discontinuities. We now show that the total amount of k-compression waves between $\chi_k^-(u_{r_i})$ and $\chi_k^-(u_{r_i})$ at time t_2 is small. This would prove the claim made at the beginning of this paragraph, at least for time t_2, by virtue of an easy generalization of Lemma 6.4. Since $\chi_k^-(u_{r_i})$ and $\chi_k^+(u_{r_i})$ are both of type I, and since waves of other families crossing them are weak, we see that they are almost straight lines. Thus we conclude that the speed $\sigma_-(t)$ and $\sigma_+(t)$ of $\chi_k^-(u_{r_i})$ and $\chi_k^+(u_{r_i})$, respectively, at time t is:

(14.4) $\sigma_-(t) = \lambda_k^- + O(1)\varepsilon$,

$$\sigma_+(t) = \lambda_k^+ + O(1)\varepsilon.$$

for certain values λ_k^- and λ_k^+ of k-characteristics speed. Thus by Lemma 9.4 we have for r_i small (or equivalently I_0 large)

(14.5) $\overline{D}(t_2) = [(\lambda_k^+-\lambda_k^-) + O(1)\varepsilon + O(r_i)] \ (t_2-t_1)$

$$= [(\lambda_k^+-\lambda_k^-) + O(1)\varepsilon] \ (t_2-t_1).$$

Now suppose that there exist (x_ℓ,t_2) and (x_r,t_2), $x_2 > x_1$, between $\chi_k^-(u_{r_i})$ and $\chi_k^+(u_{r_i})$

(14.6) $\lambda_k^r - \lambda_k^\ell = \delta < 0$

$$\lambda_k^r \equiv \lambda_k(u_{r_i}(x_r,t_2)); \qquad \lambda_k^\ell \equiv \lambda_k(u_{r_i}(x_\ell,t_2)).$$

In other words, there exist k-compression waves between $\chi_k^-(u_{r_i})$ and $\chi_k^+(u_{r_i})$ at time t_2 with an amount $|\delta|$. To show that $\delta = O(1)\varepsilon$, we may assume that (x_ℓ, t_2) and (x_r, t_2), respectively, correspond to k-characteristics $\chi_k^\ell(u_{r_i})$ and $\chi_k^r(u_{r_i})$ of type I or II. Apply the proof of Theorem 11.1 to the region between $\chi_k^-(u_{r_i})$ and $\chi_k^\ell(u_{r_i})$ and also between $\chi_k^r(u_{r_i})$ and $\chi_k^+(u_{r_i})$ to obtain

$$(14.7) \qquad \overline{D}(t_2) \geq [(\lambda_k^\ell - \lambda_k^-) + (\lambda_k^- - \lambda_k^r)] + O(1)\varepsilon](t_2 - t_1).$$

It is clear from (14.5) and (14.7) that δ in (14.8) is $O(1)\varepsilon$. This proves our claim at time t_2. We now prove the claim for time between t_1 and t_2. As noted earlier, in the region $\Omega_k^+(u_{r_i})$ not between $\chi_k^-(u_{r_i})$ and $\chi_i^+(u_{r_i})$, $u_{r_i}(x, t_2)$ consists mainly of k-compression waves. By (14.1) and conservation of k-expansion waves, (ii) of Lemma 9.2, this is so also for $u_{r_i}(x, t)$, $t_1 \leq t \leq t_2$. Thus we may apply the arguments at the end of the proof of Theorem 13.1 to treat this region. It remains to deal with the region in $\Omega_k^+(u_{r_i})$ between $\chi_k^-(u_{r_i})$ and $\chi_k^+(u_{r_i})$. We already know that in this region $u_{r_i}(x, t_2)$ consists of very weak k-compression waves. Again, by the conservation of k-compression waves and (14.1), we conclude by contradiction from (14.4) that there are no strong k-compression waves for $u_{r_i}(x, t)$, $t_1 \leq t \leq t_2$ in the region. This completes the proof of the claim made at the beginning of this paragraph.

We now complete the proof of the theorem based on the above claims. First we choose a subsequence of $\{r_i\}$ so that the sequence $\{\chi_k^1(u_{r_i}),$ $\chi_k^2(u_{r_i})\}$ tends to $\{\chi_k^1 \equiv \chi_k^1(u_0), \chi_k^2 \equiv \chi_k^2(u_0)\}$, which are Lipschitz continuous curves through (x_0, t_0). The rest of the theorem follows from the above claims. For brevity, we prove the theorem for $u_0|\Omega_k^+$. When (u_{k-1}, u_k) consists of a single k-rarefaction wave, this follows from the last claim in an obvious way. When (u_{k-1}, u_k) contains k-discontinuities $\Gamma_{k,1}, \Gamma_{k,2}, \ldots$, we know from the last claim that u_{r_i}, i large, also contains corresponding k-discontinuities. Thus we may choose a subsequence of $\{r_i\}$ so that these discontinuities in u_{r_i} tend to Lipschitz continuous curves in u_0. It is clear that these curves represent k-discontinuities in $u_0|\Omega_k^+$. This completes the proof of the theorem.

Q.E.D.

§15. Regularity of the solution

THEOREM 15.1. There exist subsets Λ_1 and Λ_2 of $\{(x,t) : -\infty < x < \infty,$ $t \geq 0\}$ with the following properties. Λ_1 consists of countable Lipschitz continuous curves and Λ_2 consists of countable points. Each curve Γ in Λ_1 represents a curve of jump discontinuity in the weak solution u_0 in the sense that except for countable points on Γ, u_0 has an admissible jump discontinuity across Γ satisfying the jump condition (R-H) as characterized in Theorem 13.1. Each point (x_0,t_0) in Λ_2 represents a point of interaction for u_0 in the sense that in a small neighborhood of (x_0,t_0), u_0 is characterized as in Theorem 14.1. Finally, outside $\Lambda_1 \cup \Lambda_2$, u_0 consists of points of continuities as characterized in Theorem 12.1.

PROOF: The theorem is a consequence of Theorems 12.1, 13.1 and 14.1. We define Λ_2 to be the set of points at which $dQ(u_0) + dC(u_0)$ has a nonzero point measure. Since $dQ(u_0) + dC(u_0)$ has finite total measure in $t \geq 0$, we see that Λ_2 consists of countable points. The set Λ_1 is constructed as follows: Let $\{t_1,t_2,\ldots\}$ be the set of nonnegative rational numbers. For each fixed t_i, $u_0(x,t_i)$ has bounded variation in x and so it is continuous except for countable values of x: $x_{i1}, x_{i2}, \ldots$. According to Theorems 13.1 and 14.1, there is a finite number of discontinuities issued from (x_{ij},t_i), $j \in \{1,2,\ldots\}$. Here we only pick those discontinuities whose initial strengths at (x_{ij},t_i) are nonzero. For each discontinuity with strength S we construct a curve Γ in Λ_1 as follows: Choose an upper half disk with center (x_{ij},t_i) with the property that the total amount of interactions and cancellations in the open half disk is less than $(\alpha S)^3$, α is a small positive number to be chosen later. We now continue the curve $\Gamma_{ij}^1,\ldots,\Gamma_{ij}^\ell$ until it reaches the upper boundary of the half disk. When the system is genuinely nonlinear, this can always be done, because the strength of a shock wave is weakened only through cancellations, and, possibly, interactions. These effects are assumed to be small and so the shock waves issued from (x_{ij},t_i) remain to have positive strength in the half disk. However, in general a discontinuity may weaken its strength by splitting into several discontinuities when it becomes nonsimple. In this case we notice that if a discontinuity of strength S is a composite of several discontinuities with strength $\{s_j\}$, then there exists α, which depends only on the system (1.1) (c.f. (3.2)), such that $\max_j\{s_j\} \geq \alpha S$. Thus we may continue the discontinuity along the discontinuities with maximum strength and so it is defined up to the upper boundary of the half disk. This completes the definition of the set Λ_1. It remains to show that any point (x_0,t_0) characterized in Theorem 13.1 is on one of the curves in Λ_1. It suffices to consider irrational t_0. For such a point there exists a disk centered at (x_0,t_0) such that the total amount of cancellations and interactions in the disk is small. By Theorem 13.1 there exist at least one, and possibly several, discontinuity of the same family through (x_0,t_0). Moreover, these discontinuities are defined up to the upper and lower boundary of the disk. Choose the one Γ with maximum strength and a point (x_1,t_1) on it, $t_1 < t_0$, such that t_1 is rational and $|t_1 - t_0|$ is small. Thus Γ has a strength at (x_1,t_1) which is larger than $|t_1 - t_0|$. Consequently, according to our construction of Λ_1, the discontinuity $\overline{\Gamma} \in \Lambda_1$ issued from (x_1,t_1), which is a part of Γ in a small neighborhood of (x_1,t_1), is indeed defined up to time $t_0 + \varepsilon$, ε positive and small, and so (x_0,t_0) belongs to $\overline{\Gamma}$ in Λ_1. This completes the proof of the theorem.   Q.E.D.

§16. Asymptotic behavior of the solution

As before we assume that the solution of the Riemann problem (u_ℓ, u_r) consists of i-waves (u_{i-1}, u_i), $i = 1,2,\ldots,n$, and (u_{i-1}, u_i) consists of i-discontinuities (i-rarefaction waves) $(u_{i,j-1}, u_{i,j})$, j odd (j even). The asymptotic state of the solution is analogous to the outgoing waves in Theorem 14.1. For this reason we only give a brief statement of the asymptotic result in the following theorem.

THEOREM 16.1. The weak solution $u_0(x,t)$ of (1.1), (1.2) tends to the solution of the Riemann problem (1.1), (1.4), $u_\ell \equiv u_0(-\infty)$, $u_r \equiv u_0(+\infty)$, in the following sense: When the i-discontinuity $(u_{i,j-1}, u_{i,j})$, j odd, is simple, there exists $T > 0$ such that a Lipschitz continuous curve $\Gamma_{i,j}$ of jump discontinuity in $u_0(x,t)$ is defined for time larger than T is and the one-sided limits of $u_0(x,t)$ along $\Gamma_{i,j}$ tend to $u_{i,j-1}$ and $u_{i,j}$ as t tends to infinity. When the i-discontinuity $(u_{i,j-1}, u_{i,j})$, j odd, is a composite of several weaker i-discontinuities, then corresponding to each weaker i-discontinuity, there exists a curve of jump discontinuity in $u_0(x,t)$ which satisfies the analogous properties as $\Gamma_{i,j}$ above. Between these curves of jump discontinuities, $u_0(x,t)$ approaches the solution of the Riemann problem (u_ℓ, u_r) uniformaly as t tends to infinity.

PROOF: Let ε be an arbitrary small positive number. Since $u_0(x,t)$ is of bounded variation in x for each t, there exists $M \equiv M(t,\varepsilon)$ such that the total variation of $u_0(x,t)$ for $|x| \geq M$ is less than ε. Since the total amount of wave interaction and cancellation is finite, there exists $T > 0$ such that the total amount in $\{(x,t) : t \geq T\}$ is less than ε. Let $\chi_k^1(u_{r_i})$ and $\chi_k^2(u_{r_i})$ be the first and last characteristics from the left of type I or II issued at time T between $(-M,T)$ and (M,T). We then proceed with the arguments in Sections 12, 13, and 14, in particular the proof of Theorem 14.1, to prove the theorem. Thus we take t_1 much larger than T so that $(t_1-T)\varepsilon > M$. In the proof of Theorems 12.1, 13.1 and 14.1 we were allowed to choose the distance between $\chi_k^1(u_{r_i})$ and $\chi_k^2(u_{r_i})$ to be much smaller than the time interval $t_2 - t_1$, because $\chi_k^1(u_{r_i})$ and

$\chi_k^2(u_{r_i})$ approaches a point we considered. In the present situation, the distance M between $\chi_k^1(u_{r_i})$ and $\chi_k^2(u_{r_i})$ at initial time $t = T$ may be large, but we are free to choose the time interval $t_1 - T$ to be arbitrarily large (c.f. (12.9)). Details are left to the reader.   Q.E.D.

The above arguments differ from those of [29] where we generalized the method of [16] to prove the theorem for genuinely nonlinear systems. The following theorem is an easy corollary of the above theorem.

THEOREM 16.2. Suppose that the initial data have compact support. Then the weak solution $u_0(x,t)$ decays uniformly to zero as t tends to infinity.

§17. Linear and nonlinear waves

Consider the equations of gas dynamics in Lagrangian coordinates:

$$\frac{\partial v}{\partial t} - \frac{\partial u}{\partial x} = 0 \quad \text{(conservation of mass)}$$

$$\frac{\partial u}{\partial t} + \frac{\partial p}{\partial x} = 0 \quad \text{(conservation of momentum)}$$

$$\frac{\partial E}{\partial t} + \frac{\partial (pu)}{\partial x} = 0 \quad \text{(conservation of energy)}$$

where u, v, e and p are, respectively, the velocity, specific volume, internal energy and the pressure of the gas and $E = e + u^2/2$ is the total energy. The pressure is usually assumed to be a known function of v and s, the entropy,

$$p = p(v,s) \quad \text{(constitutive relation).}$$

The system is strictly hyperbolic if $\partial p/\partial v < 0$:

$$\lambda_1 = -\sqrt{-\partial p/\partial v}, \qquad \lambda_2 = 0, \qquad \lambda_3 = \sqrt{-\partial p/\partial v}.$$

Thus we see that the second characteristic field is always linearly degenerate. The first and third characteristic fields are genuinely nonlinear if $\partial^2 p/\partial v^2 \neq 0$. Our weaker nonlinearity assumption in Section 3, (3.2), is satisfied for i = 1,3, if for each fixed entropy s, $\partial^2 p/\partial v^2$, as a function of v, has isolated zeros. When this is the case, the results in the previous sections hold for first and third fields. Thus, for instance, there are countable Lipschitz continuous curves of jump discontinuity pertaining first and third characteristic fields. For each such curve of discontinuity, there exists an approximate discontinuity in the approximate solution u_{r_i}, r_i sufficiently small.

Direct calculations show that the pressure p and the velocity u are constant along $R_2 = S_2$ and strictly monotone, at least locally, along R_i and S_i, $i = 1,3$. Thus the above remark shows that the pressure and velocity functions behave nonlinenarly. In particular, the discontinuity curves for these functions can be effectively calculated by random choice method. On the other hand, one can easily construct approximate solutions consisting of many weak 2-waves and yet converges to a single 2-wave in the exact solution.

REFERENCES

1. Antman, S. and Liu, T.-P., Travelling waves in hyperelastic rod, Quart. Appl. Math., 36 (1979), 377-400.

2. Bakhrarov, N., On the existence of regular solutions in the large for quasilinear hyperbolic systems, Zh. Vychisi. Mat. Mathemat. Fig., 10 (1970), 969-980.

3. Chorin, A., Random choice methods with applications to reacting gas flow, J. Comp. Phys., 25 (1977), 253-371.

4. Concus, P. and Proskurowski, W., Numerical solution of a nonlinear hyperbolic equation by the random choice method, J. Comp. Phys. pp. 153-166.

5. Conley, C. and Smoller, J., Shock waves as limits of progressive wave solutions of higher order equations, Comm. Pure Appl. Math., 24 (1971), 459-472.

6. Colon, J., A theorem in ordinary differential equations with an application to hyperbolic conservation laws, Advances in Math., 35 (1980), 1-18.

7. Conlon, J. and Liu, T.-P., Admissibility criteria for hyperbolic conservation laws (Preprint).

8. Courant, C. and Friedrichs, K. O., "Supersonic Flow and Shock Waves", Springer-Verlag, 1948.

9. Dafermos, C., Application of the invariance principle for compact processes, II. Asymptotic behavior of solutions of a hyperbolic conservation law, J. Differential Equations, 11 (1972), 416-424.

10. _____, Elasticity equations are special (Preprint).

11. DiPerna, R., Uniqueness of solutions of hyperbolic conservation laws, Indiana Univ. Math. J., 28 (1979), 137-187.

12. _____, Existence in the large for quasilinear hyperbolic conservation laws, Arch. Rat. Mech. Anal., 52 (1973), 244-257.

13. _____, Decay and asymptotic behavior of solutions to nonlinear hyperbolic systems of conservation laws, Indiana Univ. Math. J., 24 (1975), 1047-1071.

14. _____, Singularities of solutions of nonlinear hyperbolic systems of conservation laws, Arch. Rat. Mech. Anal., 60 (1975), 75-100.

15. Glimm, J., Solutions in the large for nonlinear hyperbolic systems of equations, Comm. Pure Appl. Math., 18 (1965), 695-715.

16. Glimm, J. and Lax, P., Decay of solutions of systems of nonlinear hyperbolic conservation laws, Amer. Math. Soc. Memoirs, 101 (1970).

17. Greenberg, J., On the elementary interactions for the quasilinear wave equation $\frac{\partial \gamma}{\partial t} - \frac{\partial v}{\partial x} = 0$, $\frac{\partial v}{\partial t} - \frac{\partial \sigma(\gamma)}{\partial x} = 0$, Arch. Rat. Mech. Anal., 43 (1971), 325-349,

18. John, F., Formation of singularities in one-dimensional nonlinear wave propagation, Comm. Pure Appl. Math., 27 (1974), 377-405.

19. Lax, P., Hyperbolic systems of conservation laws, II, Comm. Pure Appl. Math., 10 (1957), 537-566.

20. _____, Development of singularities of solutions of nonlinear hyperbolic partial differential equations, J. Math. Phys., 5 (1964), 611-613.

21. _____, Shock waves and entropy, "Contribution to Nonlinear Functional Analysis", ed. E. Zarantenello, 603-634, Academic Press, N.Y., (1971).

22.   Liu, T.-P., The Riemann problem for general 2 × 2 conservation laws,
      Trans. Amer. Math. Soc., 199 (1974), 89-112.

23.   _____, The Riemann problem for general system of conservation laws, J.
      Differential Equations, 18 (1975), 218-234.

24.   _____, The entropy condition and the admissibility of shocks, J. Math.
      Anal. Appl., 53 (1976), 78-88.

25.   _____, Uniqueness theorem of the Cauchy problem for general 2 × 2 con-
      servation laws, J. Differential Equations, 20 (1976), 369-388.

26.   _____, Initial-boundary value problems for gas dynamics, Arch. Rat.
      Mech. and Anal., 64 (1977), 137-168.

27.   _____, Asymptotic behavior of solutions of general system of nonlinear
      hyperbolic conservation laws, Indiana Univ. J., 27 (1978), 211-253.

28.   _____, Decay to N-waves of solutions of general system of nonlinear
      hyperbolic conservation laws, Comm. Pure Appl. Math., 30 (1977),
      585-610.

29.   _____, Large-time behavior of solutions of initial and initial-boundary
      value problem of general system of hyperbolic conservation laws, Comm.
      Math. Phys., 55 (1977), 163-177.

30.   _____, Linear and nonlinear large-time behaviors of solutions of hyper-
      bolic conservation laws, Comm. Pure Appl. Math., 30 (1977), 767-796.

31.   _____, The deterministic version of the Glimm scheme, Comm. Math.
      Phys., 57 (1977), 135-148.

32.   _____, Development of singularities in the nonlinear waves for quasi-
      linear hyperbolic partial differential equations, J. Differential
      Equations, 33 (1979), 92-111.

33.   Nishida, T., Global solution for an initial boundary value problem of
      a quasilinear hyperbolic system, Proc. Jap. Acad., 44 (1968), 642-646.

34.   Nishida, T. and Smoller, J., Solutions in the large for some nonlinear
      hyperbolic conservation laws, Comm. Pure Appl. Math., 26 (1973),
      183-200.

35.   Oleinik, O., On the uniqueness of the generalized solution of a Cauchy
      problem for a nonlinear system of equations occurring in mechanics,
      Uspehi Mat. Nauk., 73 (1957), 169-176 (in Russian).

Department of Mathematics
University of Maryland
College Park, MD 20742

ABCDEFGHIJ—AMS—8987654321